CONTENTS

THE LIVING EARTH

ASTEROIDS, COMETS, AND METEORITES

COSMIC INVADERS OF THE EARTH

JON ERICKSON

FOREWORD BY TIMOTHY KUSKY, Ph.D.

® Checkmark Books®
An imprint of Facts On File, Inc.

ASTEROIDS, COMETS, AND METEORITES
Cosmic Invaders of the Earth

Checkmark Books
An imprint of Facts On File, Inc.
132 West 31st Street
New York NY 10001

Library of Congress Cataloging-in-Publication Data

Erickson, Jon, 1948–
    Asteroids, comets, and meteorites : cosmic invaders of the earth / Jon Erickson.
        p. cm. — (The living earth)
Includes bibliographical references and index.
    ISBN 0-8160-4873-8 (hardcover: alk paper)   ISBN 0-8160-5076-7 (pbk)
    1. Asteroids. 2. Comets. 3. Meteorites. I. Title.
    QB651 .E75 2003
    551.3'97—dc21   2002002434

Checkmark Books are available at special discounts when purchased in bulk quantities for
businesses, associations, institutions, or sales promotions. Please call our Special Sales
Department in New York at 212/967-8800 or 800/322-8755.

You can find Facts On File on the World Wide Web at http://www.factsonfile.com

Text design by Cathy Rincon
Cover design by Nora Wertz
Illustrations by Jeremy Eagle © Facts On File

Printed in the United States of America

VB Hermitage 10 9 8 7 6 5 4 3 2 1

This book is printed on acid-free paper.

TABLES

ACKNOWLEDGMENTS

The author thanks the Defense Nuclear Agency, the National Aeronautics and Space Administration (NASA), the National Oceanic and Atmospheric Administration (NOAA), the National Optical Astronomy Observatories (NOAO), the U.S. Air Force, the U.S. Department of Energy, the U.S. Geological Survey (USGS), and the U.S. Navy for providing photographs for this book.

The author also thanks Frank K. Darmstadt, Senior Editor, and the staff of Facts On File for their invaluable contributions to the making of this book.

FOREWORD

Asteroids, comets, and meteorites have been objects of fascination, speculation, and fear for most of recorded human history. Early peoples thought that fiery streaks in the sky were omens of ill fortune and sought refuge from their evil powers. Impacts of comets and meteorites with Earth are now recognized as the main cause for several periods of mass extinction on the planet, including termination of the dinosaurs 66 million years ago. Comets and meteorites may also have brought much of the water, air, and perhaps even life to Earth. The planets themselves coalesced from numerous smaller asteroids, comets, and interplanetary dust. Asteroids, comets, and meteorites were therefore essential for the formation of Earth and life and were also responsible for the end of the line for many species.

In this book, Jon Erickson brilliantly presents the reader with a fascinating and readable treatise on asteroids, comets, and meteorites. The book starts with a discussion on the origin of the solar system, Sun, and planets and then looks at the role of these planetsimals in the formation of Earth. Erickson then examines the importance of impacting meteorite and comets in the heavy bombardment period that marks the first 500 million years of Earth's history. This is followed by a survey of different impact craters recognized on other planets in the solar system.

Erickson proceeds with an inventory of the different asteroid belts. He includes descriptions of and historical accounts about the discovery of some of the larger asteroids like the 600-mile wide Ceres, discovered in 1801 by the Italian astronomer Giuseppe Piazzi. The origins of comets in the Kuiper belt and distant Oort cloud is presented in lucid terms and leads one to wonder about the possibility that the building blocks of life came from these far reaches of the solar system. If life came from out there, are people truly alone?

The final few chapters of the book deal with the possibility of large meteorites and comets hitting Earth and the consequences of such collisions. On June 30, 1908, a huge explosion rocked a sparsely populated area of central Siberia. Scientists now believe this explosion was caused by a fragment of comet Encke that broke off the main body as it passed Earth. Early in the morning of June 30, a huge fireball moved westward across the Siberian sky. Then an explosion centered near Tunguska was so powerful that it knocked people off their feet hundreds of miles away. A 12-mile-high fireball was visible for 400 miles, and trees were charred and knocked down in a 2,000 square kilometer area.

Earth has experienced even larger impacts. Several have caused 50 to 90 percent of all species alive on the planet at the time of the impact to go extinct, paving the way for the evolution and diversification of new organisms. The impact of a 6-mile-wide meteorite with the Yucatán Peninsula is now thought to have ended the reign of the dinosaurs. The impact instantly formed a fireball 1,200 miles across, followed by a tsunami hundreds if not thousands of feet tall. The dust thrown out of the deep crater excavated by the impact plunged the world into a fiery darkness and then months or even years of freezing temperatures. As soon as the dust settled, carbon dioxide released by the impact caused Earth to soar into an intense greenhouse warming. Few species handled these changing environmental stresses well, and 65 percent of all species went extinct.

After discussing the history and consequences of impacts on Earth, Jon Erickson outlines some of the possible defenses the human race might mount against any asteroids or comets determined to be on an impact course with Earth. The future of the human race may well depend on increased awareness of how to handle this potential threat. In only 1996, an asteroid about 0.25 miles across nearly missed hitting Earth, speeding past at a distance about equal to the distance to the Moon. The sobering reality of this near collision is that the asteroid was not even spotted until a few days before it sped past Earth. In June 2002, another asteroid came within 75,000 miles of Earth. What if the object were bigger or slightly closer? Would it have been stoppable? If not, what would have been the consequences of its collision with Earth?

— Timothy M. Kusky, Ph.D.

INTRODUCTION

The science of meteoritics deals with the study of meteorites and their impacts on Earth. Meteorite craters on the Moon, the inner planets, and the moons of the outer planets are quite evident and numerous. Several remnants of ancient meteorite craters remain on Earth, suggesting it was just as heavily bombarded as the rest of the solar system. Meteorite impacts have produced many strikingly circular features in the crust scattered throughout the world. They are testimony to disasters in the past caused by major meteorite impacts. In the future, many more craters will be found, painting a clear picture of what transpired long ago. The evidence hints that impact cratering is an ongoing process, and the planet can expect another major meteorite impact at any time.

Large meteorite impacts had a major effect on the history of life since the very beginning. Throughout geologic history, asteroids and comets have repeatedly bombarded Earth, implying such events are a continuing process. Sometimes asteroids the size of mountains struck the planet, extinguishing large numbers of species. The most celebrated extinction was the death of the dinosaurs and many other species. The dinosaur killer left its footprints all over the world.

Between the orbits of Mars and Jupiter lies a wide belt of asteroids containing a million or more stony and metallic irregular rock fragments. Several large asteroids have been observed to be out of the main asteroid belt and in Earth-crossing orbits. Comets strike at much higher velocities, making them more deadly. Many Earth-bound asteroids and comets capable of killing millions of people are wandering around in space. Dozens of near-Earth asteroids have been discovered, and occasionally a wayward object wanders near out planet.

The text begins with the origin of the universe, the galaxies, and the solar system. It then examines the creation of Earth, the Moon, and the initiation of life. The text continues with historical meteorite impact events on Earth and explores craters on the other planets and their moons. Next it discusses asteroids, the asteroid belt, meteors, and meteorites, comets, and meteor showers. The text continues with an examination of meteorite craters and impact structures scattered around the world. It then discusses the global effects of large meteorite impacts and the impact theory of mass extinction of species. Finally, it examines the effects to civilization of large asteroid or comet impacts on Earth.

Science enthusiasts will particularly enjoy this fascinating subject and gain a better understanding of how the forces of nature operate on Earth. Students of geology and Earth science will also find this a valuable reference to further their studies. Readers will enjoy this clear and easily readable text that is well illustrated with photographs, illustrations, and helpful tables. A comprehensive glossary is provided to define difficult terms, and a bibliography lists references for further reading. Meteorite impacts are among the tireless geologic processes that continually shape this living Earth.

1

ORIGIN OF THE SOLAR SYSTEM

FORMATION OF THE SUN AND PLANETS

Three basic theories have been put forth to explain the creation of the universe (Fig. 1): the big bang theory, in which the universe keeps expanding; the steady state theory, in which the universe exists indefinitely; and the pulsation theory, in which the universe begins and ends repeatedly. According to the big bang theory discussed here, all matter in the universe, including a wide assortment of stars, galaxies, galaxy clusters, and superclusters, was created during a gigantic explosion around 15 billion years ago. Subsequently, every chemical element in the periodic table formed during little bangs by huge exploding stars called supernovas. Consequently, we are products of the universe. Every atom in our bodies and all the materials that make up Earth came from the stars.

About two-thirds outward from the center of the Milky Way galaxy, which theoretically contains a black hole at its core—where matter and energy disappear as though down a cosmic drain pipe, lies a lone, ordinary star. That star happens to be our Sun (Fig. 2). Single, medium-sized stars, such as the Sun, are somewhat of a rarity in this galaxy. They are perhaps the only ones with orbiting planets. Thus, of the myriad of stars overhead, only a handful possess planets, and fewer yet contain life.

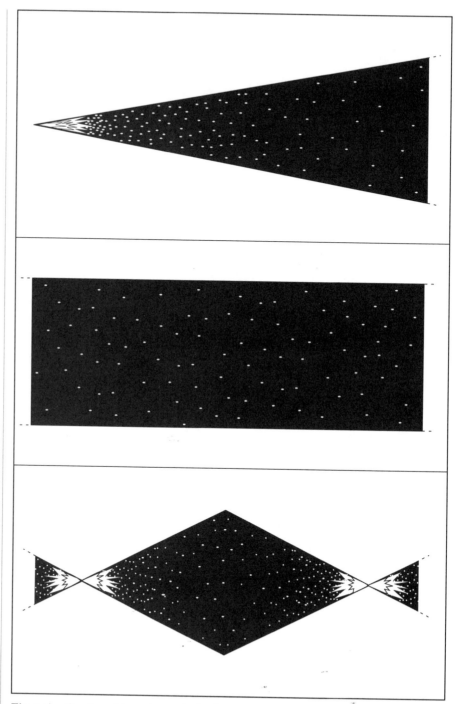

Figure 1 *Creation of the universe: the big bang theory* (top), *the steady state theory* (middle), *and the pulsation theory* (bottom).

Figure 2 *The Sun showing sunspots and a solar storm in the upper right.*

(Photo courtesy NASA)

THE BIG BANG

According to theory, the universe originated with such an explosive force that presently the farthest galaxies are hurling away at nearly the speed of light. Today, the light astronomers see from the most distant galaxies originated when the universe was less than one-fifth its present age. The beginning universe did not grow at a steady rate but might have suddenly expanded temporarily by a process called inflation. During this time, gravity might have briefly become a repulsive force rather than one of attraction, causing the cosmos to undergo a tremendous spurt of growth. The infant universe rapidly ballooned outward for an instant and then settled down to a more constant rate of expansion as it progressed to the more conventional type of development observed today. The inflation theory explains some of the universe's fundamental features, such as the uniformity of the microwave afterglow of the big bang and the apparent lack of curvature, or flatness, of space.

This great firestorm lasted about 100,000 years and involved practically all matter in the universe. Huge swirling masses of high-temperature plasma composed of elementary particles flew outward into space in all directions. Currents and eddies flowed violently through this primordial soup, forcing material to clump together. The temperature of the protouniverse eventually cooled sufficiently to allow the formation of protons and neutrons, which came together into atomic nuclei.

At some point following the big bang, fluctuations in the smooth flow of matter and energy served as seeds for galaxy formation. A change in the fabric of space produced lumps and ripples in the distribution of matter. These yielded galaxies and galaxy clusters with up to hundreds of members. This phase transition apparently occurred during the first million years or so of the universe's life after electrons had combined with protons to produce hydrogen atoms, the starting materials of the universe. Hydrogen and helium comprise more than 99 percent of all matter in the universe. Helium is continuously being generated in the stars. However, hydrogen was made only once, at the beginning of the universe, and no new hydrogen has been created since the big bang.

The universe is estimated to be composed of about 75 percent hydrogen, 25 percent helium, and minor amounts of other elements. A puzzling question concerning the universe is this abundance of helium. The gas is made up of a nucleus of two protons and two neutrons orbited by two electrons. It is observed on the surface of the Sun and was actually discovered on the Sun before it was found on Earth. Helium exists in the stars of this galaxy, in the stars of other galaxies, and in interstellar space.

The nuclear fusion reactions that power the stars and convert hydrogen into helium can account for only a minor percentage of the helium. Therefore, the bulk of the helium must have originated during the big bang. As the protouniverse continued to expand, the basic units of matter began to agglomerate into some 50 billion galaxies, each with tens of billions or hundreds of billions of stars. The universe is dominated by small- to medium-sized stars with less than 80 percent of the mass of the Sun.

Remnants of the big bang can still be found by measuring the temperature of the universe. Besides starlight, the universe radiates other forms of energy. One of these is microwave radiation spread evenly throughout the universe. Its discovery in the mid-1960s was instrumental in the development of the big bang theory. This leftover energy from the creation of the universe has presently cooled to within a few degrees above absolute zero (−273 degrees Celsius), the temperature at which all molecular motion ceases. Faint temperature fluctuations in the background microwave radiation might signify primordial lumps that later gave rise to the present-day galaxies.

The galaxies are arranged in four basic types: elliptical, spiral, irregular, and diffuse. The elliptical galaxies are quite old and spheroidal in shape, with the highest light intensity at their centers. Fully formed elliptical galaxies, which took shape over a period of 1 billion years, already existed when the universe was only one-tenth its current age, when spiral galaxies were still forming. Powerful radio sources originate most often from elliptical galaxies. Their red color suggests that these galaxies contain an abundance of old stars.

A spiral galaxy (Fig. 3), which includes the Milky Way, has a pronounced bulge at its center, much like a mini-elliptical galaxy. A spiral-patterned disk populated with young stars surrounds this bulge. The spiral arms generate magnetic fields produced by the rotation of the galaxy. Irregular galaxies, as their name implies, have many shapes and are relatively low in mass. Diffuse galaxies have low surface brightness with more gas and much less of a spiral structure, suggesting they are not fully developed.

The age of the universe is determined by measuring the distance and speed of the farthest known galaxies some 15 billion light-years from Earth by using the red shift of their starlight. The color of light emitted by stars shifts to the longer wavelengths, or the red end of the electromagnetic spectrum (Fig. 4), when the star is moving away. The farthest galaxies therefore have the largest red shifts, signifying they are traveling the fastest. A paradox seems to exist, however, because the universe appears to be younger than its oldest stars due to uncertainties about the value of the Hubble constant used for measuring the rate of expansion of the universe.

Astronomers glimpse back toward the very beginning of the universe by observing an apparent massive protogalaxy in its formative stages. It is 12 billion light-years from Earth, meaning the object was seen as it existed only a few billion years after the big bang. In the meantime, the Milky Way galaxy (Fig. 5), which is of fairly modest size, pulled enough matter together to form a large spiral galaxy, similar to many others observed in the far reaches of space. The Milky Way contains some 100 billion stars spread out across a distance of about 100,000 light-years.

Astronomers can also weigh the universe to determine whether it will continue to expand, collapse upon itself into a dense cosmic soup, or remain in a steady state, with new galaxies forming to fill the voids created by the expansion. The weight of the universe signifies its gravitational attraction. It can be determined by measuring the mass of an average galaxy and multiplying that number by the total number of galaxies.

Yet more matter appears to exist than that observed in the visible universe. This is known as the *missing mass*. This unseen dark matter might be many times more massive than all the stars combined and could contain up to 90 percent of all the mass in the universe. Moreover, the supposed dark matter in the Milky Way's halo, a large region that extends well beyond the galaxy's visible outline, is what holds the rapidly rotating galaxy together. At least half this missing matter could reside in ordinary dead stars called white dwarfs, objects as small as Earth but 1 million times denser. Without large amounts of hidden mass, the galaxy would simply fly apart. Furthermore, not knowing how much mass is

Figure 3 *The Whirlpool galaxy.*
(Photo courtesy NASA)

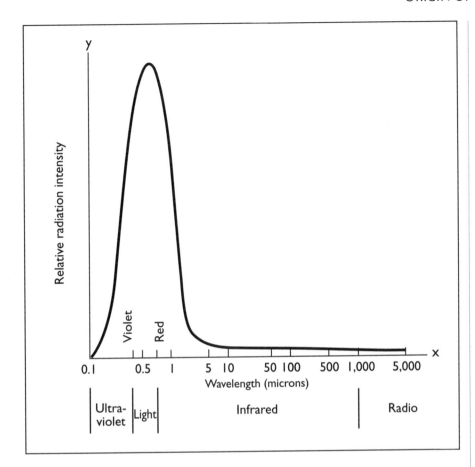

Figure 4 *The solar spectrum.*

missing means that the ending of the universe, whether it expands forever or collapses in on itself, will remain a perplexing enigma.

GALAXY FORMATION

The earliest galaxies evolved within the first billion years of the creation of the universe, after it had expanded to about one-tenth its present size. Elliptical galaxies had been in existence for some time, while spiral galaxies such as the Milky Way were just forming. Strong gravitational or electromagnetic forces exerted by cosmic strings, which are vast concentrations of energy and the largest known structures in the universe, gathered matter around them to initiate galaxy formation. Even the smallest of these thin clouds or ripples stretch across 500 million light-years of space. The largest known structure in the universe, the *great wall,* is a collection of galaxies within a few hundred million light years of Earth. It forms a sheet some 300 million light-years long.

The galaxies combined into clusters, which accumulated into superclusters that wandered through space in seemingly haphazard motions. Superclusters are filamentary in shape and extend up to hundreds of millions of light-years across. This makes them some of the largest structures. Most massive stars either collapsed into black holes or exploded and became supernovas, providing the raw materials for new stars (Fig. 6). The stars created all the known chemical elements, which later became the building blocks for new stars and planets, including the Sun and Earth.

Stars appear to be the most common components of galaxies. Nonetheless, astronomers have found strange objects called compact structures in the interstellar medium that might prove to be as much as 1,000 times more

Figure 5 *The Milky Way galaxy.*

(Photo courtesy NOAO)

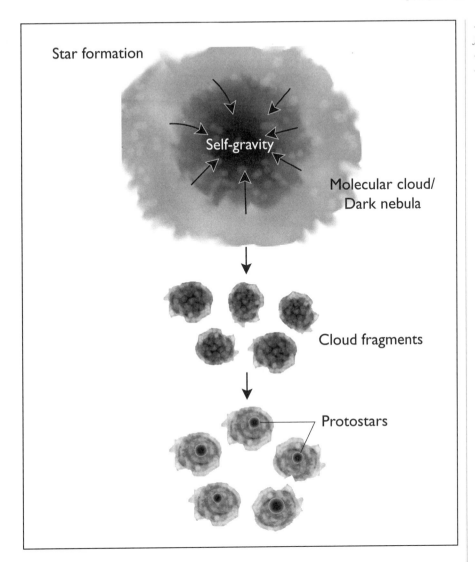

Star formation

Self-gravity

Molecular cloud/
Dark nebula

Cloud fragments

Protostars

Figure 6 Star formation from the collapse of the solar nebula by a nearby supernova.

numerous. These objects are roughly spherical blobs of ionized matter and measure about as wide as Earth's orbit around the Sun. They are therefore larger than all but the biggest stars.

The presence of compact structures is revealed only when they block radio waves emanating from distant quasars, which occupy the centers of hyperactive young galaxies that were much more common in the past than they are today. Quasars are the most luminous bodies in the universe. They burn intensely for about 100 million years before dying out. They are so far away that the light presently seen from them originated when the universe was only one-tenth of its present age and one-quarter of its current size. When the compact structures move in front of these powerful radio wave sources, the radio signals cease. The objects remain in this position for a month or more. When they move out of the way, the radio signals return. They move too fast to reside outside this galaxy, otherwise they would have to be traveling several times the speed of light.

Globular clusters are great aggregations that pack upward of 1 million stars in a small volume of space. They are the oldest groupings of stars in the Milky Way, reaching 11 billion years old. These structures were found to have a high density in the constellation of Sagittarius (Fig. 7). Radio astronomers pinpointed this region as the source of intense radio emissions generated by the interaction of electrons and protons in clouds of ionized gas. The clouds might mark the location of a massive, collapsed object such as a black hole or a compact cluster of hot, young stars at the heart of the galaxy.

Figure 7 *The Swan Nebula located in the constellation Sagittarius.* (Photo courtesy NASA)

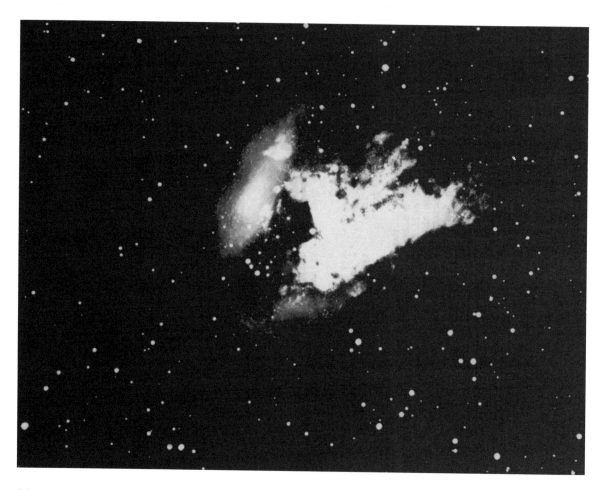

The black hole probably formed simultaneously with the central bulge of the galaxy as gas migrated to the core. Powerful gamma rays and X rays traveling at the speed of light also radiate from this zone. The concentration of energy suggests that Sagittarius is the center of the galaxy. Other high-energy cosmic and gamma ray sources must lie outside the Milky Way because the explosions that emitted them would have destroyed part of the galaxy.

The Milky Way galaxy has a central bulge with a radius of about 15,000 light-years consisting of densely packed old stars. Five spiral arms peel off the central bulge somewhat like the swirling clouds of a hurricane. New stars are born in these dense regions from interstellar gas and dust particles. The galaxy revolves once around the central core roughly every 250 million years.

Outside the central bulge lies the galactic disk, which has a radius of 50,000 light-years, wherein this solar system resides. It comprises relatively young stars along with large quantities of gas and dust. The mass of the visible stars appears to account for practically all material in the galactic disk. New stars originate in dense regions of interstellar matter called giant molecular clouds. A galactic halo with a radius of 65,000 light-years has widely spaced old stars along with roughly half the globular clusters in the galaxy. It also contains large amounts of dark matter, which provides the gravitational forces to hold the galaxy together.

The dark matter in the Milky Way apparently comprises upward of 90 percent of the galactic mass. Most galaxies rotate so fast they would fly apart if their visible stars provided the only source of gravity. The stars in the outer reaches of the galaxy move as fast as those nearer the center, which suggests the galaxy's mass is spread out and not concentrated in the galactic core.

Beyond the galactic halo lies the outermost region of the galaxy, called the corona. It stretches out to about 300,000 light-years from the galactic center. The corona contains old burned-out stars, brown dwarfs, and poorly luminous objects called galactic companions. These include globular clusters, dwarf spheroidal galaxies, and the irregular galaxies that make up the Large and Small Magellanic Clouds. The Large Magellanic Cloud galaxy, which lies about 180,000 light-years from Earth, is a conspicuous patch of light near the south celestial pole.

The Andromeda galaxy (Fig. 8), about 2.3 million light-years from Earth, is the closest large spiral galaxy and resembles in many respects the Milky Way. Both galaxies appear to be a binary galactic system and orbit each other around a common center of gravity. The Milky Way is believed to be following an orbit in the shape of a narrow, flat ellipse that presently takes it away from Andromeda. Perhaps in about 4 billion years, both galaxies could experience a close encounter with each other. Even if a collision did occur, the distance between stars is so vast that little disruption would be expected.

Figure 8 *The Andromeda galaxy, 2 million light-years from Earth.*

(Photo courtesy NASA)

STELLAR EVOLUTION

Once or twice a century, in some part of the galaxy, a giant star more than 100 times larger than ordinary stars explodes in a miniaturized big bang, forming a supernova 1 billion times brighter than the Sun. The outer matter of the star is flung out into space at fantastic speeds. Large amounts of radiation are also released, producing deadly cosmic rays. The energetic particles bombard Earth's atmosphere and collide with other particles in the air, sending them showering to the ground.

Supernovas are thought to play an enormous role in the formation of new stars. The gigantic exploding stars burn extremely hot, giving them life spans of only a few hundred million years. The stars that produce supernovas also create all the known elements. This occurs first by fusing hydrogen into helium, then fusing helium into the lighter elements such as carbon and oxygen, and finally fusing these into the heavy elements on up to iron.

Energy flowing out from the fusion reaction at the core prevents the star from collapsing in on itself by its own gravitation. However, when the core turns to iron, no further fusion reactions are possible, and the star begins to collapse. The center of the core is crushed to a density of hundreds of millions of times greater than normal matter, somewhat like shrinking Earth down to about the size of a golf ball. The shock wave created by material falling into the core causes matter to rebound out into space, producing the rapid brightening of a supernova. When the star explodes, this matter is expelled into empty space, and the gas and dust are later swept up by newly forming stars and their planets.

When a star reaches the supernova stage, after a very hot existence spanning several hundred million years, the nuclear reactions in its core become an explosive event. The star sheds its outer covering, while the core compresses to an extremely dense, hot body called a neutron star. The expanding stellar matter from the supernova forms a nebula (Fig. 9) composed mostly of

Figure 9   *The solar system condensed from a dense cloud similar to the Orion Nebula.*

(Photo courtesy NASA)

hydrogen and helium along with particulate matter that comprises all the other known chemical elements.

The shock wave from a nearby supernova compresses portions of the nebula, causing the nebular matter to collapse upon itself by gravitational forces into a protostar. As the nebula continues to collapse, it begins to gyrate, flattening into a pancake-shaped disk. Spirals of matter segregate into concentric rings, which eventually coalesce into planets. Meanwhile, the compressional heat initiates a thermonuclear reaction in the core, and a star is born.

A new star is created in the Milky Way galaxy every few years or so. Stars originate from an assortment of nebulas, molecular complexes, and globules composed of condensing clouds of gas and dust. A typical globule has a radius of about one light-year and a density of about 100 solar masses. Along with abundant hydrogen and helium, the galactic clouds also contain an assortment of organic compounds, including carbon monoxide, formaldehyde, and ammonia.

As the center of the solar nebula collapses into a dense glowing ball of hydrogen and helium, the intense heat generated by the gravitational compression initiates a thermonuclear reaction in the core. Some 10 million years after the nebula first began to collapse, the star ignites. If the star is rather large, its strong particle radiation, called the T-Tauri wind, blows away all the surrounding gas and dust before this matter can coalesce into planets, leaving behind a lone giant.

Multiple star systems of two or more stars form when the central nodule fissions. They are quite common and comprise about 80 percent of all the stars in the galaxy. The stars orbit each other around a common center of gravity. For the most part, they are not believed to possess planets due to strong gravitational interactions that interfere with planetary formation.

A rapidly spinning star would also not be expected to have planets—at least not for very long. Even if planets had formed, they would be unable to hold their orbits and would spiral into the star due to its large gravity. Fortunately, for the planets in this solar system, nearly all of the angular momentum, or rotational energy, resides with the planets, which keeps them in their respective orbits. The nearly circular orbits of most planets in the solar system appear to be exceptional because extrasolar planets generally have elongated, egg-shaped paths.

SOLAR ORIGINS

About 4.6 billion years ago, the Sun, an ordinary main-sequence star, was created in one of the dusty spiral arms of the Milky Way galaxy. A density wave emanating from a nearby supernova provided the outside pressure needed to

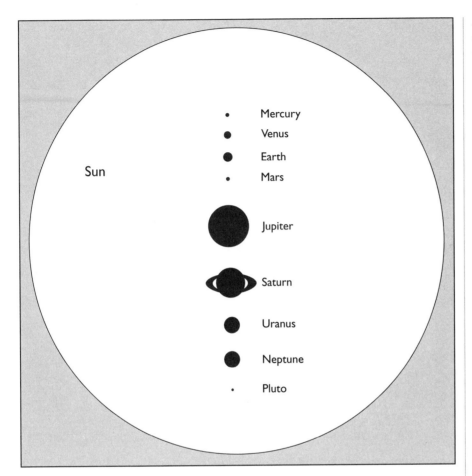

Figure 10 The approximate relative sizes of the Sun and planets.

Sun

Mercury
Venus
Earth
Mars
Jupiter
Saturn
Uranus
Neptune
Pluto

start the collapse of interstellar matter by self-gravitation. As the solar nebula continued to collapse, it rotated faster and faster, while spiral arms peeled off the rapidly spinning nebula to form a protoplanetary disk.

Orbiting the Sun is a collection of planets (Fig. 10 and Table 1), their moons, large rocks from broken-up planetoids called asteroids, and comets formed from leftover gases and ices swept out of the solar system by the early Sun's strong solar wind. When the Sun first ignited, it produced intense particle radiation. The radiation pressure created a solar wind that was more like a gale when compared with the gentle breeze of today. The strong solar wind possibly stripped away the volatiles of the inner planets and deposited them further out in the solar system, providing material for the larger gaseous planets.

Debris from a nearby supernova mingling with the original gas and dust was incorporated in the construction of the solar system. Late arrivals condensed independently into a primitive class of meteorites called carbonaceous

TABLE 1   SUMMARY OF SOLAR SYSTEM DATA

Body	Orbit in millions of miles	Radius miles	Mass	Density	Axis tilt	Rotation	Year	Temp. (°C)	Atmospheric composition
Mercury	36	1,500	0.1	5.1	10	58.6 days	88 days	425	Carbon dioxide
Venus	67	3,760	0.8	5.3	6	242.9 days	225 days	425	Carbon dioxide minor water
Earth	93	3,960	1.0	5.5	23.5	24 hours	365 days		78% nitrogen 21% oxygen
Mars	141	2,110	0.1	3.9	25.2	24.5 hours	687 days	−42	Carbon dioxide minor water
Jupiter	483	44,350	318	1.3	3.1	9.9 hours	11.9 years	2000	60% hydrogen 36% helium 3% neon 1% methane and ammonia
Saturn	886	37,500	95	0.7	26.7	10.2 hours	29.5 years	2000	Same as Jupiter
Uranus	1,783	14,500	14	1.6	98	10.8 hours	84 years		Similar to Jupiter, no ammonia
Neptune	2,793	14,450	18	2.3	29	15.7 hours	165 years		Same as Uranus
Pluto	3,666	1,800	0.1	1.5		6.4 days	248 years		

chondrites, which are alien in composition to the rest of the solar system. Much of this leftover debris revolves around the Sun in erratic orbits that can take them across Earth's path.

During the Sun's first billion years, it was extremely unstable. The solar output was only about 70 percent of its present intensity. As a result, the Sun provided only as much warmth on Earth as it presently does on Mars. Periodically, powerful nuclear reactions in the Sun's core created large thermal pressures that caused the Sun to expand up to one-third larger than its present size. The internal turmoil produced enormous solar flares that leaped millions of miles into space. A high-temperature plasma of atomic particles created a strong solar wind. This intense activity made the Sun radiate more heat. This, in turn, cooled the core, allowing the Sun to return to its normal size.

The early Sun spun rapidly on its axis, completing a single rotation in just a few days as compared with 27 days at present. This rapid rotation produced strong magnetic fields similar to how a generator produces more electricity the faster it turns. The strong magnetic fields helped slow the Sun's rapid rotation and transferred its angular momentum to the orbiting planets. The magnetic fields also created considerable turmoil on the Sun's surface, producing numerous giant sunspots and solar flares much greater than those seen today (Fig. 11).

The Sun consists of an inner core of helium, an intermediate layer of hydrogen called the radiative zone, and an outer layer of hot gases called the

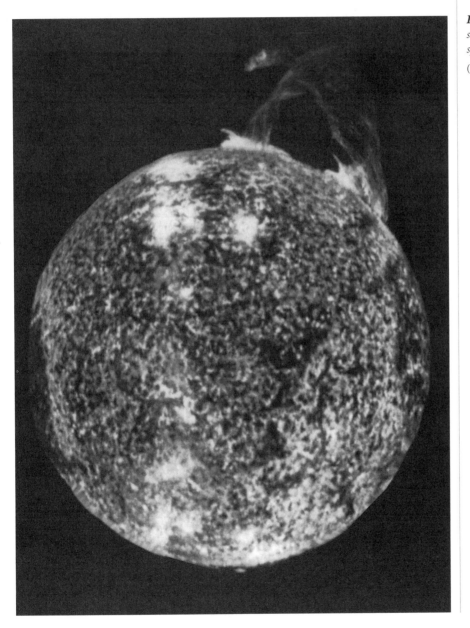

Figure 11   *The Sun showing one of the most spectacular solar flares.*

(Photo courtesy NASA)

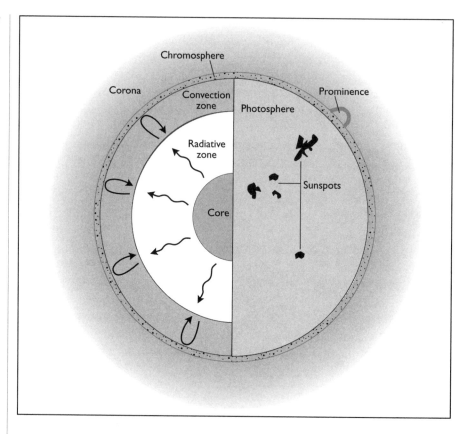

convection zone (Fig. 12). The photosphere, the visible surface of the Sun, has a temperature of about 6,000 degrees Celsius. It is often marred by sunspots, which are large patches of relatively cool gas associated with magnetic vortices and vigorous solar flares. Sunspots are used to determine the rate of the Sun's rotation since they are the only landmarks on an otherwise bare surface. Energetic solar particles shoot out from the Sun's equator and bombard Earth with intense particle radiation.

The chromosphere is the Sun's atmosphere. It is composed of reddish glowing gases thousands of miles thick. An intensely hot halo called the corona extends millions of miles into space and is visible only during a total eclipse, when the Moon moves directly in front of the Sun. The surface of the Sun is furiously roiling. Jets of gas spurt millions of miles high, arch over, and fall back to the surface. Globules of white-hot gas rise to the surface, cool, and descend again. This boiling layer extends downward as far as one-third the distance toward the center of the Sun and can reach temperatures of several millions of degrees. The region is still not sufficiently hot to support nuclear reactions, which occur only in the core.

The Sun is expected to continue shining normally for another 5 billion years. The solar wind is now so feeble that it will not affect the Sun's rotation rate significantly. Violent activities, such as solar flares, will probably become less pronounced. The Sun will continue to heat up as it depletes the hydrogen fuel in the core, which becomes polluted with helium "ash."

In about 1.5 billion years, the Sun's luminosity, or brightness, will increase about 15 percent. On Earth, the ice caps will melt. This will cause massive flooding as the northern regions become hot deserts. Life-forms will have to evolve mechanisms to compensate for the higher temperatures to survive. If humans are still around, they will have to modify their environment to deal with the heat. Perhaps cities would be covered with air-conditioned domes, similar to modern football stadiums.

By the time the Sun is 10 billion years old, the nuclear fires in the core will convert virtually all hydrogen to helium and will begin to die down. In its last dying breaths, the Sun will balloon outward 40 percent larger than its present radius and glow twice as brightly. Meanwhile, all life on Earth will become extinct.

In another 1.5 billion years, as the last of the hydrogen fuel is consumed, the core will gradually contract due to a lack of pressure to resist the weight of the overlying layers. The rise in gravitational contraction will increase the temperature of the core, eventually igniting the hydrogen fuel in the overlying regions. As this fuel burns, the Sun will continue expanding until it reaches the present orbit of Mercury. It will then become a red giant with a surface temperature of only half its present value but with a luminosity 500 times greater. In the meantime, Earth will be completely incinerated.

The Sun will live a comparatively short life as a red giant, perhaps only 250 million years. Afterward, the nuclear reaction in the core will begin to convert helium into carbon and oxygen. This reaction will be so rapid it will become an explosive event known as a *helium flash*. The outer layers will be blown off, removing a significant portion of the Sun's mass while hurling all the planets into interstellar space. The Sun will then contract to about 10 times its present diameter and settle down to a steady burning of helium.

As the Sun again gradually grows, successive outer layers will be blown off by the intense solar wind until the core is finally exposed, resulting in a ring nebula with debris surrounding a very hot stellar remnant in the center (Fig. 13). Eventually, the debris will clear away, unveiling a lone hot star with only about half its previous mass and compressed down to about the size of Earth. Its extreme heat will cause it to glow white hot, hence the name white dwarf. The Sun will live on in this manner for perhaps another 15 billion years. Then it will gradually cool down to end its days as a cold black dwarf.

PLANETARY ASSEMBLY

During the Sun's early stages of development, it was ringed by a protoplanetary disk composed of several bands of coarse particles about the size of gravel called planetesimals (Fig. 14). They originated from primordial dust grains created by a supernova that clumped together by weak electrical and gravitational attractions.

Figure 13   *The Ring Nebula in Lyra.*

(Photo courtesy NOAO)

When the Sun first ignited, the strong solar wind blew away the lighter components of the solar nebula and deposited them in the outer regions of the solar system. The remaining planetesimals in the inner solar system comprised mostly stony and metallic fragments, ranging in size from fine sand grains to huge boulders. In the outer solar system, where temperatures were

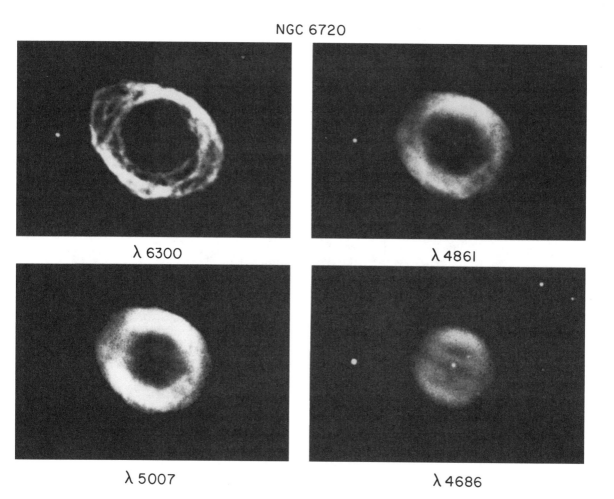

NGC 6720

λ 6300

λ 4861

λ 5007

λ 4686

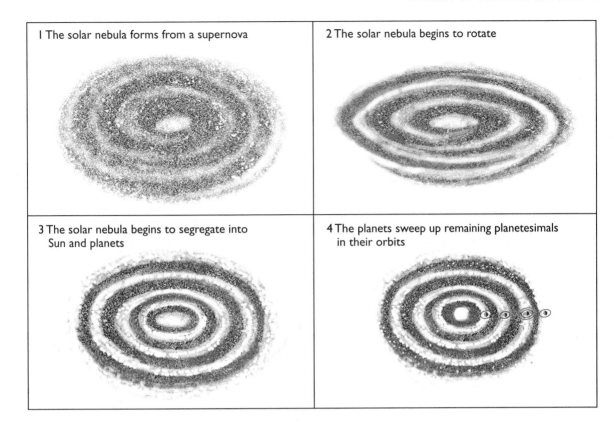

I The solar nebula forms from a supernova

2 The solar nebula begins to rotate

3 The solar nebula begins to segregate into Sun and planets

4 The planets sweep up remaining planetesimals in their orbits

much colder, rocky material along with solid chunks of water ice, frozen carbon dioxide, and crystalline methane and ammonia condensed.

The outer planets are believed to possess rocky cores about the size of Earth, a mantle possibly composed of water ice and frozen methane, and a thick layer of compressed gas, mostly hydrogen and helium along with smaller amounts of methane and ammonia. Pluto, the moons of the outer planets, and the comets are essentially rocks encased in a thick layer of ice or are composed of a jumble of rock and ice. The composition of Jupiter is similar to that of the Sun. Had it continued growing, Jupiter might have become sufficiently hot to ignite into a small brown dwarf star, resulting in a twin star system similar to many in this galaxy.

Up to 100 trillion planetesimals orbited the Sun during the solar system's early stages of development. As they continued to grow, the small rocky chunks swung around the infant Sun in highly elliptical orbits along the same plane, called the ecliptic. The constant collisions among planetesimals formed larger bodies. After about 10,000 years of formation, some planetary bodies

Figure 14   *The solar nebula segregated into concentric rings of planetesimals, which coalesced into the planets.*

Figure 15   *Saturn from* Voyager 1 *in November 1980.*

(Photo courtesy NASA)

grew to more than 50 miles wide. However, most of the planetary mass resided with the small planetesimals.

    If not for the presence of a large gaseous medium in the solar nebula to slow the planetesimals, the larger bodies would have continued to sweep up the remaining planetesimals. This would have resulted in a solar system composed of thousands of planetoids roughly 500 miles in diameter, about the size of the largest asteroids in existence today. The outcome would have been a solar system resembling the planet Saturn and its rings (Fig. 15).

    An incredible amount of water, one of the simplest of molecules, resides in the solar system. As the Sun emerged from gas and dust, tiny bits of ice and rock debris began to gather in a frigid, flattened disk of planetesimals surrounding the infant star. The temperatures in some parts of the disk might have been warm enough for liquid water to exist on the first solid bodies in the solar system. In addition, water vapor in the primordial atmospheres of the inner terrestrial planets might have eroded away by planetesimal bombardment and blown beyond Mars by the strong solar wind of the infant Sun. Once planted in the far reaches of the solar system, the water coalesced to create icy bodies that streak by the Sun as comets.

The solar system is quite large, consisting of nine known planets and their moons (Fig. 16). The image of the original solar disk can be traced by observing the motions of the planets. All of them revolve around the Sun in the same direction it rotates. In fact, all except Pluto do so within 3 degrees of the orbital plane or ecliptic. Because Pluto's orbit inclines 17 degrees to the ecliptic, it might be a captured planet or possibly a moon of Uranus knocked out of orbit by collision with a comet. Another Uranian satellite, named Miranda (Fig. 17), appears to have been blasted apart by an impact and reassembled with all its

Figure 16 The solar system, showing the orbits of the planets and their relative sizes and distances.

(Photo courtesy NASA)

THE MILKY WAY GALAXY

ORBITS OF THE PLANETS

THE EARTH AND MOON

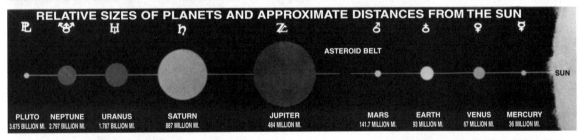
RELATIVE SIZES OF PLANETS AND APPROXIMATE DISTANCES FROM THE SUN

ASTEROID BELT

SUN

PLUTO	NEPTUNE	URANUS	SATURN	JUPITER	MARS	EARTH	VENUS	MERCURY
3.675 BILLION MI.	2.797 BILLION MI.	1.787 BILLION MI.	887 MILLION MI.	484 MILLION MI.	141.7 MILLION MI.	93 MILLION MI.	67 MILLION MI.	36 MILLION MI.

MOON

EARTH

VENUS

MERCURY   SUN SPOTS

MARS

SOLAR PROMINENCE

SATURN

JUPITER

URANUS

NEPTUNE

PLUTO

THE SOLAR SYSTEM
AS SEEN LOOKING TOWARD EARTH FROM THE MOON

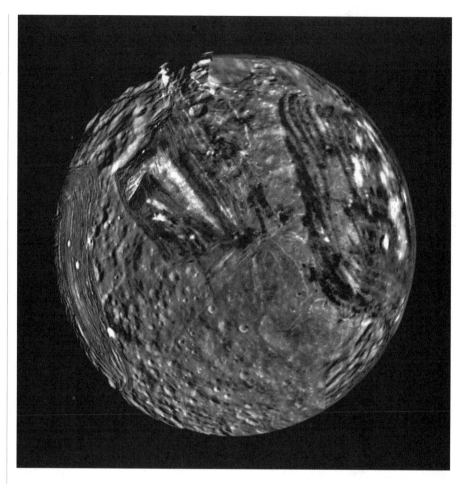

pieces in the wrong places. A small unseen planet, named Vulcan for the Roman god of fire, is predicted to exist inside the orbit of Mercury.

Some 7 billion miles from the Sun lies the heliopause, which marks the boundary between the Sun's domain and interstellar space. About 20 billion miles from the Sun is a region of gas and dust, possibly remnants of the original solar nebula. Several trillion miles from the Sun is a shell of comets that formed from the leftover gas and ice of the original solar nebula. Some comets break loose, possibly by the gravitational attraction of a passing star, and enter the inner solar system, sometimes passing very close to Earth.

After discussing the origin of the universe, the galaxies, and the solar system, the next chapter examines the creation of Earth, the Moon, and the initiation of life.

2

THE FORMATION
OF EARTH

PLANETARY ORIGINS

This chapter examines the origin of this planet, the Moon, and life. Of all the planets and the 60 some satellites in the solar system, Earth is unique with its roving continents and evolving life-forms. It is a middle-aged planet about halfway through its life cycle. The age of Earth is based on the ages of meteorites formed about the same time as the rest of the solar system, some 4.6 billion years ago. This date agrees well with the age of Earth's oldest rocks, which formed about 4.2 billion years ago.

In its early stages of formation, a wave of giant impactors pounded the infant Earth. As many as three Mars-sized bodies are thought to have struck the planet, one of which might have created the Moon. Several remnants of ancient craters not completely erased by Earth's active geology suggest that Earth was as heavily bombarded as the rest of the planets and their moons. The meteorite impacts might also have removed Earth's primordial atmosphere and initiated the formation of continents, the presence of which are unmatched anywhere in the solar system. Numerous giant meteorites slamming into Earth added unique ingredients to the boiling cauldron, leading to the initiation of life.

THE BIG CRASH

Early in the formation of the solar system, the Sun was ringed by a rotating protoplanetary disk composed of several bands of coarse particles called planetesimals. These grew by the attraction of primordial dust grains as they traveled in highly eccentric orbits around the Sun. The ones closest to the Sun comprised stony and metallic minerals, ranging in size from sand grains to huge bodies more than 50 miles across. Those farther away contained mainly volatiles and gases blown outward by the violent solar winds of the early Sun.

Around 4.6 billion years ago, as the Sun's first fiery rays blazed across the heavens, the primal Earth emerged from a spinning, turbulent cloud of gas, dust, and planetoids surrounding the infant star. During the next 700 million years, the debris cloud settled down into a more tranquil solar system, and the third planet from the Sun began to take shape. The core and mantle segregated possibly within the first 100 million years. During that time, Earth was in a molten state produced by radioactive isotopes and impact friction from planetesimals. The presence of magnetic rocks as old as 3.5 billion years suggests that Earth at an early age had a molten outer core and a solid inner core comparable to their present sizes.

Earth's metallic core might have formed first by mutual magnetic attraction of metallic planetesimals that originally comprised the cores of early planetoids that disintegrated after impacts with other bodies. The study of metallic meteorites, which once formed the cores of early planetoids that have disintegrated, suggests that Earth's core is composed of iron and nickel. When the core was nearly fully developed, gravitational forces were strong enough to attract the stony planetesimals, which built up layer by layer above the core. The completed process of accretion took about 100 million years. It resulted in a planet segregated into a molten metallic core, a plastic mantle, and a rocky crust (Fig. 18). Earth's original crust has long since disappeared, remixed into the interior by the impact of giant meteorites that were leftovers from the creation of the solar system.

Earth could also have formed unsegregated out of a homogeneous mixture of rock and metal derived from captured planetesimals. The infalling planetesimals heated the surface by impact friction, while radioactive elements warmed the planet from the inside out. Earth melted and segregated into concentric layers. The heavier metallic elements sank toward the center, and the lighter materials rose toward the surface. The core attracted siderophilic, or iron-loving, materials such as gold, platinum, and certain other elements originating from meteorite bombardments during the planet's early formation. The entire process of segregating the core and mantle took place within the first 100 million years, resulting in a planet that was molten throughout.

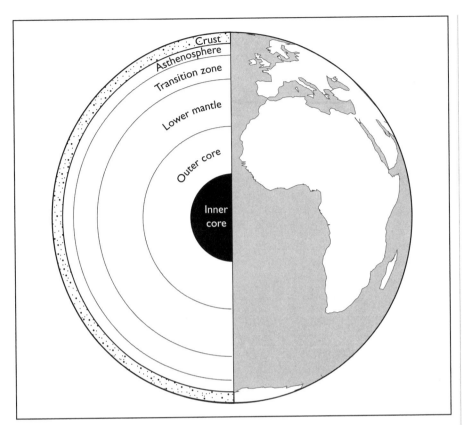

The core comprises a solid inner sphere composed of iron-nickel silicates surrounded by an outer layer of liquid iron. This two-part structure along with Earth's high rotation rate generates electrical currents that produce a strong magnetic field. The geomagnetic field creates a magnetosphere, which envelops the planet far out in space and protects it from deadly cosmic radiation originating from the Sun and other parts of the galaxy.

Earth continued growing by accumulating planetesimals with temperatures often exceeding 1,000 degrees Celsius. As the red-hot orb evolved, its orbit began to decay due to drag forces created by leftover gases in interplanetary space. The formative planet slowly spiraled inward toward the Sun, sweeping up additional planetesimals along the way. Eventually, Earth's path around the Sun was completely swept clean of interplanetary material, leaving a gap in the disk of planetesimals, while its orbit stabilized near its present position.

In the early stages, Earth's interior was hotter, less viscous, and more vigorous. Heavy turbulence in the mantle, whose heat flow was three times greater than at present, produced violent agitation on the surface. This turmoil created a sea of molten and semimolten rock broken up by giant fissures, from which huge fountains of lava spewed skyward.

During the first half-billion years, Earth's surface was scorching hot. Heat of compression from the primordial atmosphere with pressures 100 times greater than that today resulted in surface temperatures hot enough to melt rocks. When the Sun ignited, the strong solar wind blew away the lighter components of Earth's atmosphere. A massive bombardment of meteorites blasted the remaining gases into space, leaving the planet in a near vacuum much like the Moon is today.

In the absence of an atmosphere to hold in Earth's internally generated heat, the surface rapidly cooled, forming a primitive crust similar to the slag on molten iron ore. The thin basaltic crust was comparable to that presently on Venus (Fig. 19). Indeed, Earth's Moon and the inner planets offer clues as to the planet's early history. Among the features common to the terrestrial planets were their ability to produce voluminous amounts of basaltic lavas.

Figure 19 *Radar images of Venus's northern latitudes, from Russian Venera spacecraft.*

(Photo courtesy NASA)

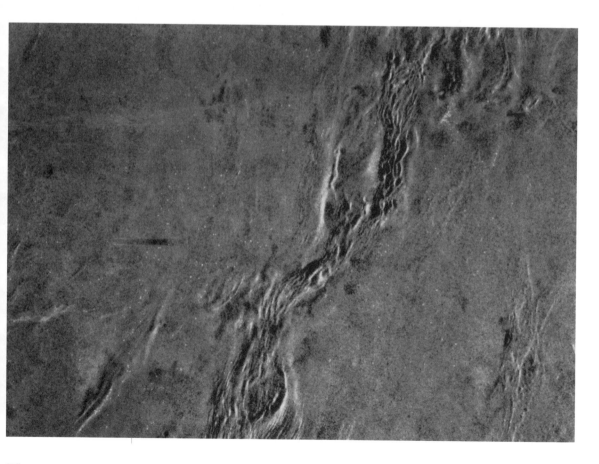

This hardened basaltic layer was not a true crust, however, because the interior of Earth was still in a highly molten and agitated state. Strong convective currents stirred the mantle and kept it well mixed, thus preventing chemical separation of the lighter and heavier constituents. Therefore, the density of the rocks solidifying on the surface was equivalent to those of the mantle. The highly unstable crust remelted on the surface, or sank into the mantle and remelted, or grew top-heavy, overturned, and remelted. Earth's original crust remixed into the interior by strong convection and by the impacts of giant meteorites.

As the meteorites plunged into the planet's thin basaltic crust, they gouged out huge quantities of partially solidified and molten rock. The scars in the crust quickly healed as batches of fresh magma bled through giant fissures and poured onto the surface, creating a magma ocean. Because the early crust was so unstable, no geologic record exists for the first 700 million years, known as the Hadean eon. For this reason, no impact structures older than about 4 billion years have been found.

The formative Earth was subjected to intense volcanism and meteorite bombardment that repeatedly destroyed any crust attempting to solidify. This explains why the first half-billion years of Earth's history are missing from the geologic record. Because Earth had not yet formed a permanent crust, the impactors simply plunged into the hot planet.

When the crust began to form, a massive meteorite bombardment melted large portions by impact friction. The magma formed lava seas, which solidified into lava plains similar to the maria on the Moon. Much of the crust became large impact basins, with walls rising nearly 2 miles above the surrounding terrain and floors plunging to 10 miles in depth.

The original crust was quite distinct from modern continental crust, which first appeared more than 4 billion years ago and represents less than 0.5 percent of the planet's total volume. During this time, Earth spun wildly on its axis, completing a single rotation every 14 hours. The high rotation rate likely arose as the planet was forming by the accretion of planetesimals. The impactors imparted angular momentum to Earth, causing it to spin faster and faster. The high rotational energy maintained high temperatures throughout the planet.

Present-day plate tectonics (Fig. 20), the interaction of large blocks of crust, could not have operated under these hot conditions, which resulted in more vertical bubbling than horizontal sliding. Therefore, modern-style plate tectonic processes were probably not fully functional until around 2.7 billion years ago, when the formation of the crust was nearly complete. Ironically, because of plate tectonics, with its sinking oceanic plates into Earth's interior, no craters on the ocean floor are older than 200 million years.

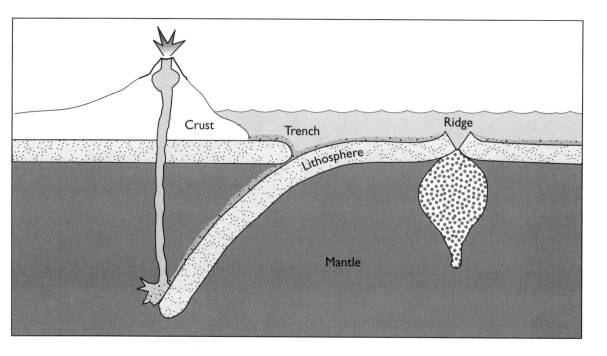

Figure 20 The plate tectonics model in which lithosphere created at midocean ridges is subducted into the mantle at deep-sea trenches, causing the continents to drift around Earth.

Earth has the thinnest crust of all the terrestrial planets (Fig. 21). Information about the early crust is provided by ancient rocks that have survived intact over the eons. They formed deep within the crust a few hundred million years after the formation of the planet and presently outcrop at the surface. Zircon crystals (Fig. 22) found in ancient granitic rocks are enormously resistant and tell of the earliest history of Earth, when the crust first formed some 4.2 billion years ago.

Among the oldest rocks are those of the 4-billion-year-old Acasta gneiss, a metamorphosed granite in the Northwest Territories of Canada. Their existence suggests that the formation of the crust was well underway by this time. The discovery suggests that at least small patches of continental crust existed on Earth's surface at an early date. Earth apparently took less than half its history to form an equivalent volume of continental rock as it has today.

After the formation of the crust, a massive meteorite shower involving thousands of impactors, many as wide as 50 miles or more, bombarded Earth and the Moon between 4.2 and 3.9 billion years ago. A swarm of debris left over from the creation of the solar system bombarded Earth. The bombardment might have delivered heat and organic compounds to the planet, sparking the rapid formation of primitive life. Alternatively, the pummeling could have wiped out existing life-forms in a colossal mass extinction.

The bombardment eventually tapered off to a somewhat steady rain of asteroids and comets. When Earth acquired a permanent atmosphere and ocean, intense weather systems eroded all impact craters. As a result, no telltale signs of the great bombardment can be found.

THE BIG SPLASH

A compelling theory for the creation of the Moon contends that a large celestial body collided with the early Earth when it was much smaller and still in a molten state. While the planet was forming out of bits of metal and rock, a planetoid about the size of Mars that was wandering out in space was

Figure 21 Comparison of topographies of Earth, Mars, and Venus.

(Photo courtesy NASA)

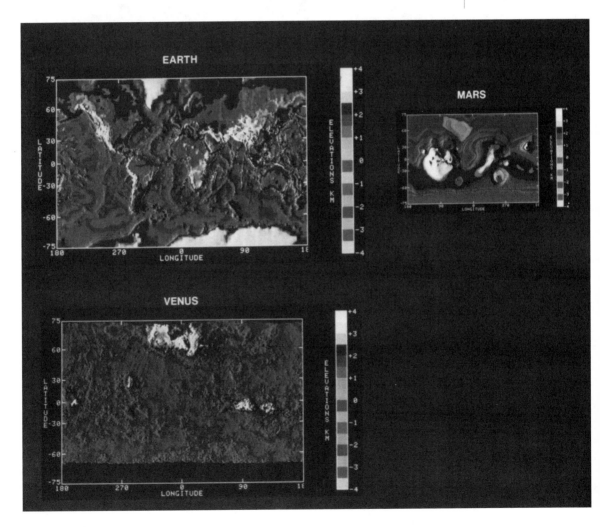

31

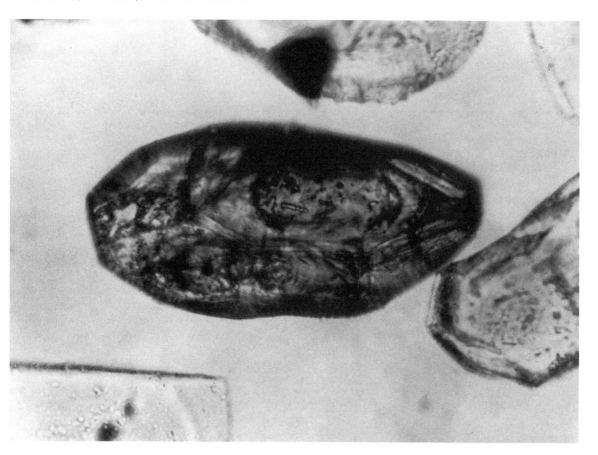

Figure 22 *Zircons from
the rare-earth zone, Jasper
Cuts area, Gilpin
County, Colorado.*

(Photo by E. J. Young,
courtesy USGS)

knocked out of its orbit around the Sun by Jupiter's strong gravitational
attraction or by a collision with a wayward comet. On its way toward the
inner solar system, the object glanced off Earth somewhat like a cosmic bil-
liard ball (Fig. 23).

The tangential collision, spanning a period of about half an hour, created a powerful explosion equivalent to the detonation of a quantity of
dynamite equal to the mass of the planetoid itself. The collision tore a great
gash in Earth. A large portion of its molten interior splashed into orbit
around the planet, forming a ring of debris called a prelunar disk. Much of
the debris from the collision rained down on Earth, splashing into an ocean
of molten rock.

The impact might also have knocked Earth over, tilting its rotational axis
and giving the planet its four seasons (Fig. 24). Similar collisions between large
bodies and the other planets, especially Uranus, which orbits on its side like a

bowling ball, might explain their various degrees of tilt and elliptical orbits. All planets except Uranus and Venus also rotate in the same direction. Their high rotation rates were possibly influenced by large asteroid bombardments.

The glancing blow might also have increased Earth's angular momentum, speeding up the planet's rotation rate from a single turn once every week to a rate considerably faster than today. The high rotation rate would have generated temperatures high enough to melt the entire planet. Unlike earlier models of lunar formation such as fission, capture, or assembly in place, the theory better explains the present rotational speed of the Earth–Moon system.

Earth's new satellite continued growing as it swept up rocky material in orbit around the planet. Infalling bodies in orbit around the Moon also crashed to the surface. Eventually, the Moon cleaned away all debris in its eccentric path around Earth. Meanwhile, the Moon heated up by radioactivity, compression, and impact friction, becoming a molten daughter planet. The Moon became gravitationally locked onto Earth, rotating at the same rate as the satellite's orbital period, always keeping the same side facing its mother

Figure 23 *The giant impact hypothesis of lunar formation envisions a Mars-sized planetesimal (A) colliding with the proto-Earth, resulting in a gigantic explosion and the jetting outward of both planetesimal and proto-Earth material into orbit around the planet (B). A proto-Moon begins to form from a prelunar disk (C), and matter accretes to form the Moon (D).*

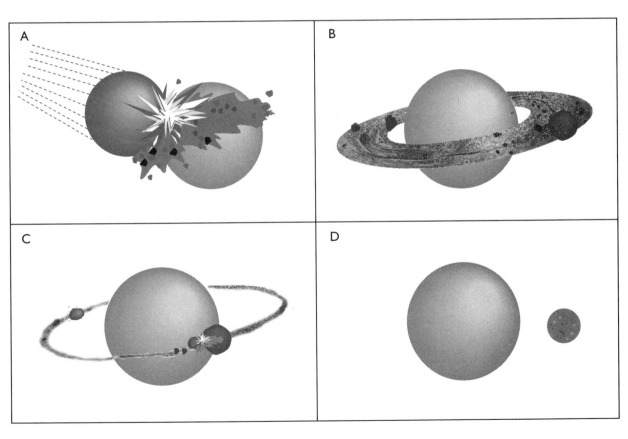

planet at all times. Many moons in orbit around other planets share this characteristic, suggesting they formed in a similar manner.

The massive meteorite shower that bombarded Earth pounded the Moon as well. Numerous large asteroids struck the lunar surface and broke through the thin crust. Huge rock fragments in orbit around the Moon also crashed onto the surface, which explains the presence of many large impact craters seen on the lunar terrain. Great floods of dark basaltic lava spilled onto the surface, providing the Moon with a landscape of giant craters and flat lava plains (Fig. 25).

Since Earth's sister planet Venus formed in a similar fashion and is equal in many respects, the absence of a Venusian moon is quite curious. It might have crashed into its mother planet or escaped into orbit around the Sun. Comparable in size and composition with Earth's Moon is the planet Mercury (Fig. 26), which perhaps was once a moon of Venus.

With a much smaller mass and volume, the Moon quickly cooled and formed a thick crust long before Earth did. Convection in the Moon's mantle and molten iron core might have generated a weak magnetic field. How-

Figure 24 *The effect of the tilt of Earth's axis on incoming solar energy, which determines the seasons.*

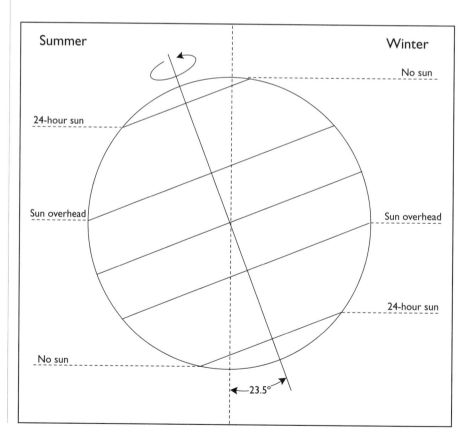

Figure 25 *Apollo spacecraft photograph of an earthrise over the lunar horizon.*

(Photo courtesy Earthquake Information Bulletin 52, USGS)

ever, convective motions were not strong enough to drive crustal plates as they do on Earth. During the great meteorite bombardment, the Moon was highly cratered. However, because it lacked an atmosphere to erode the surface, it retains most of the original terrain features.

Evidence for this theory of lunar origin was obtained from Moon rocks (Fig. 27) brought back during the Apollo missions of the late 1960s and early 1970s. The rocks appeared to be similar in composition to Earth's upper mantle and range in age from 3.2 to 4.5 billion years old. Since no rocks were found any younger, the Moon probably ceased volcanic activity, and the interior began to cool and solidify at an early age. In addition, the Moon's relatively tiny core roughly 500 miles in diameter or about 2 percent of the lunar mass, as compared with Earth's core of about 30 percent of the planet's mass, indicates the Moon was born with a severe iron deficiency. This feature suggests that the Moon was made of relatively iron-poor mantle material torn out of Earth by a giant impact.

After its formation, the Moon was in such a low orbit it filled much of the sky. The newly formed Moon circled just 14,000 miles above Earth, racing around the planet every two hours. Its strong gravitational attraction created

Figure 26   *A simulated encounter with Mercury by* Mariner 10 *in March 1974.*

(Photo courtesy NASA)

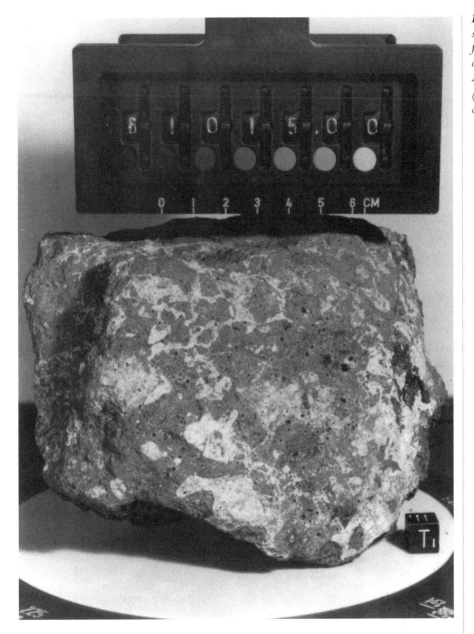

huge tidal bulges in Earth's thin crust. The presence of a rather large Moon, the most massive in relation to its mother planet of all the moons in the solar system, together with Earth forming a double planetary system, might have had a major influence on the initiation of life. The unique properties of the Earth-Moon system raised tides in the ocean (Fig. 28). Cycles of wetting and drying

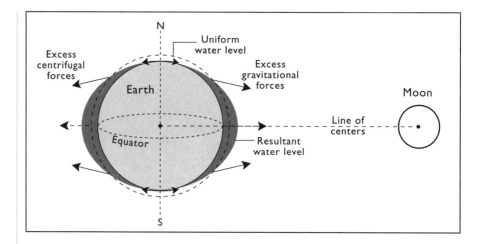

in tidal pools might have spurred the evolution of life much earlier than previously thought possible.

Furthermore, the Moon might have been responsible for the relatively stable climate, making Earth hospitable to life by stabilizing the tilt of the planet's rotational axis, which marks the seasons. Without the Moon, life on Earth would likely face the same type of wild fluctuations in climate that Mars seems to have experienced through the eons. With its spin axis no longer maintained by the Moon, in only a few million years Earth would have drastically altered its tilt enough to make the polar regions warmer than the tropics.

Earth rotated on its axis much more rapidly than it does at present, and the days were only a few hours long. As late as 900 million years ago, Earth spun 30 percent faster, and a day lasted 18 hours (Fig. 29). The Moon exerts a force on the spinning Earth called nutation, which causes the rotational axis to precess or wobble like a toy top. Tidal friction, the energy loss as the Moon causes water to slosh around the globe, slowed Earth's rotation. As time progressed, the Moon spiraled outward in an ever-widening orbit. Eventually, the Moon's orbit gradually widened to 240,000 miles. Even now, the Moon is still drifting away from Earth at about 1.5 inches a year, and the days continue to grow slightly longer.

THE BIG BURP

As the Sun emerged from gas and dust, tiny bits of ice and rock debris gathered in a frigid, flattened disk of planetesimals surrounding the infant star within the first 100 million years. Temperatures in some parts of the early solar system might have been sufficiently warm for liquid water to form on the first solid bodies. In addition, water vapor in the primordial atmospheres of the

inner terrestrial planets might have eroded away by planetesimal bombardment and been blown beyond Mars by the strong solar wind of the infant Sun.
Once planted in the far reaches of the solar system, ice crystals coalesced to
create comets that returned to Earth to supply it with water.

Early in Earth's formation, a barrage of asteroids and comets pounded
the infant planet and the Moon (Fig. 30). Some impactors were stony comprising rock and metal, others were icy with frozen gases and water ice, and
many contained abundant carbon. Perhaps these carbon-rich meteorites bore
the organic molecules from which life originated. Type I carbonaceous chondrites appear to be fragments of asteroid-sized parent bodies. The carbonates
they contain might represent sediments deposited by running water. Therefore, the existence of liquid water at such an early age might have hastened
the creation of life on Earth.

Comets comprising rock debris and ice also plunged into Earth, releasing tremendous quantities of water vapor and gases. The degassing of these
cosmic invaders liberated mostly carbon dioxide, ammonia, and methane,
major constituents of the early atmosphere, which began to form about 4.4

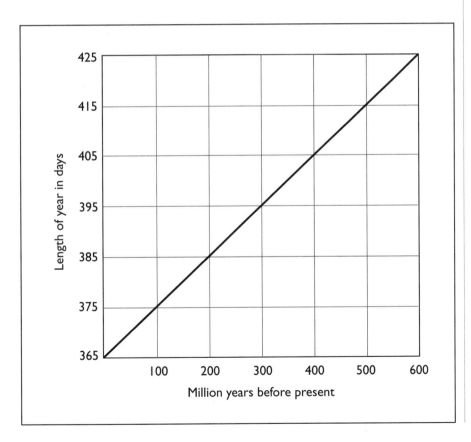

Figure 29 *The change in the length of day throughout time.*

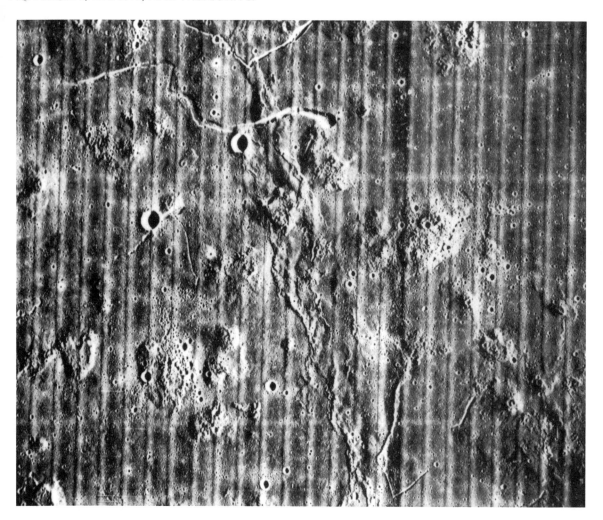

Figure 30 *The heavily
cratered Marius Hills
region on the Moon,
showing domes, ridges, and
rills on the lunar surface.*

(Photo by D. H. Scott,
courtesy USGS)

billion years ago. The comets came from the outer reaches of the solar system,
where they formed from volatiles blown outward by the strong solar wind of
the infant Sun. Thus, some volatiles lost when Earth's original atmosphere
blew away might have been returned by way of comets.

Soon after the massive meteorite bombardment began, Earth acquired a
primordial atmosphere composed of carbon dioxide, nitrogen, water vapor,
and other gases belched from a profusion of volcanoes as well as gases such as
ammonia and methane delivered by a fusillade of comets. These latter gasses
were in sufficient quantities to provide organic matter that could have been
used to initiate life.

Most of the water vapor and gases originated from within Earth itself by
volcanic outgassing. Early in the formation of Earth, giant volcanoes erupted

huge quantities of steam and gases during a period called the *big burp*. About 80 percent of the atmosphere outgassed in the first few hundred million years, with the rest released gradually during the next 4 billion years. The early volcanoes were extremely explosive because Earth's interior was hotter and the magma contained more volatiles consisting of water and various gases.

The volcanic outgassing and meteorite degassing created a thick, steamy atmosphere with massive clouds covering the entire planet much like they presently do on Venus (Fig. 31). Indeed, Venus, with its heavy carbon dioxide atmosphere, is used as a model to describe Earth's early years. The air was so

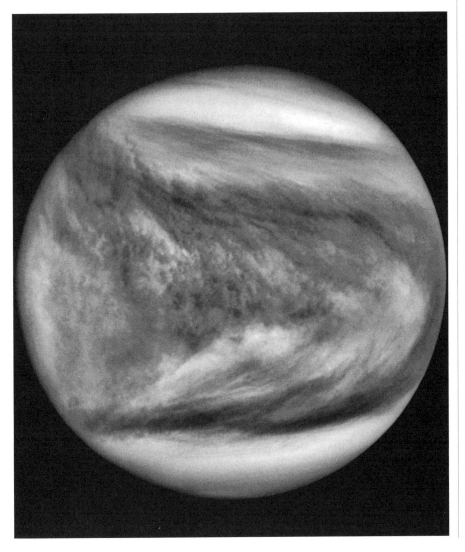

Figure 31 *Venus, from the* Pioneer Venus Orbiter *in December 1980.*

(Photo courtesy NASA)

saturated with water vapor that surface pressures were more than 100 atmospheres (atmospheric pressure at sea level). The early atmosphere contained as much as 1,000 times the current level of carbon dioxide. This was fortunate because the Sun's output was only about 70 percent of its present value, and a strong greenhouse effect kept Earth from freezing solid. The planet also retained its warmth by a high rotation rate and by the absence of continents to block the flow of ocean currents.

The primordial atmosphere contained little oxygen. Oxygen originated directly from volcanoes and meteorites and indirectly from the breakdown of water vapor and carbon dioxide by the strong ultraviolet radiation from the Sun. Oxygen generated in this manner quickly bonded to metals in the crust and also recombined with hydrogen and carbon monoxide to reconstitute water vapor and carbon dioxide. A small amount of oxygen might have lingered in the upper atmosphere, where it provided a thin ozone screen. The ozone layer blocked solar ultraviolet rays, which break down water molecules, thereby preventing the loss of the ocean, a fate that might have befallen Venus eons ago.

Nitrogen originated from volcanic eruptions and from the breakdown of ammonia, a molecule of one nitrogen atom and three hydrogen atoms and a major constituent of the primordial atmosphere. Unlike most other gases, which have been replaced or recycled, Earth retains much of its original nitrogen. This is because nitrogen readily transforms into nitrate, which easily dissolves in the ocean, where denitrifying bacteria return the nitrate-nitrogen to its gaseous state. Therefore, without life, Earth would have long ago lost its nitrogen and atmospheric pressure would be only a fraction of its present value.

THE BIG FLOOD

During the formation of the atmosphere, winds blew with a tornadic force, producing fierce dust storms on the dry surface. The entire planet was blanketed with suspended sediment similar to today's Martian dust storms, which coated the surface with windblown sediment (Fig. 32). Huge lightning bolts darted back and forth. Earth-shattering thunder sent gigantic shock waves reverberating through the air. Volcanoes erupted in one giant outburst after another.

The restless Earth rent apart as massive quakes cracked open the thin crust, and huge batches of magma bled through the fissures. Voluminous lava flows flooded the surface, forming flat, featureless plains dotted with towering volcanoes. The intense volcanism lofted massive quantities of volcanic debris into the atmosphere, giving the sky an eerie red glow. Millions of tons of volcanic debris spilled into the atmosphere and remained suspended for long

Figure 32 *The landscape of Mars, from the* Pathfinder *lander, showing boulders surrounded by windblown sediment.*

(Photo courtesy NASA)

periods. By shading the Sun, the dense dust and ash cooled Earth and provided the nuclei upon which water vapor coalesced.

As the temperatures in the upper atmosphere fell, water vapor condensed into massive clouds. The clouds were so thick and heavy they completely blocked out the Sun, plunging the surface into near darkness and dropping temperatures even further. With a continued cooling of the atmosphere, huge raindrops descended from the sky. The rains fell in torrents, producing the greatest floods the planet has ever known.

Deep meteorite craters filled rapidly like huge bowls of water, which spilled over onto flat lava plains. Raging floods cascaded down steep mountain slopes and the sides of large meteorite craters, gouging out deep canyons in the rocky plain. Giant valleys were carved out as water rushed down the steep sides of tall volcanoes, which spewed steam and gases into the atmosphere. Meanwhile, icy comets continued to plunge into the planet, releasing their load of water.

When the skies finally cleared as the thick cloud layer dissipated, Earth was transformed into a glistening blue orb covered almost entirely by a deep global ocean dotted with numerous volcanic islands. Countless volcanoes erupted undersea. Hydrothermal vents spewed hot water containing sulfur and other chemicals. Some volcanoes poked above the sea to become scattered islands, yet no continents graced Earth's watery face.

Ancient metamorphosed marine sediments of the Isua Formation in southwestern Greenland (Fig. 33) support this scenario for the creation of the ocean. These sediments are among the oldest rocks, dating to about 3.8 billion years ago, and indicate that the planet had surface water by this time. The rocks originated in volcanic island arcs and therefore lend credence to the idea that plate tectonics operated early in the history of Earth. The seawater was carried deep into Earth's interior by sinking plates and returned by volcanoes, which supplied a high mineral content, giving the ocean its high salinity.

From the end of the great meteorite bombardment to the formation of the first sedimentary rocks, vast quantities of water flooded Earth's surface. Seawater became salty due to the abundance of chlorine and sodium supplied by volcanoes. However, the ocean did not reach its present concentration of salts until about 500 million years ago. The warm ocean was heated from

Figure 33 Location of the Isua Formation in southwestern Greenland, which contains some of the oldest rocks on Earth.

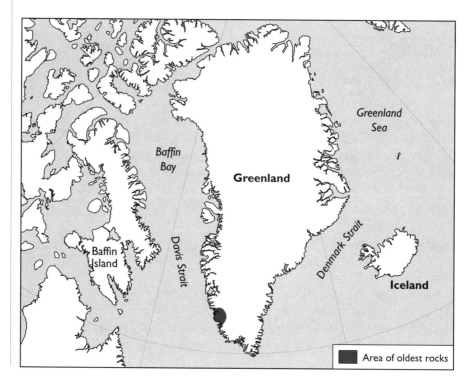

Greenland Sea

Baffin Bay

Greenland

Davis Strait

Baffin Island

Denmark Strait

Iceland

Area of oldest rocks

TABLE 2   EVOLUTION OF LIFE AND THE ATMOSPHERE

Evolution	Origin (millions of years)	Atmosphere
Humans	2	Nitrogen, oxygen
Mammals	200	Nitrogen, oxygen
Land animals	350	Nitrogen, oxygen
Land plants	400	Nitrogen, oxygen
Metazoans	700	Nitrogen, oxygen
Sexual reproduction	1,100	Nitrogen, oxygen, carbon dioxide
Eukaryotic cells	1,400	Nitrogen, carbon dioxide, oxygen
Photosynthesis	2,300	Nitrogen, carbon dioxide, oxygen
Origin of life	3,800	Nitrogen, methane, carbon dioxide
Origin of Earth	4,600	Hydrogen, helium

above by the Sun and from below by active volcanoes on the ocean floor, which continually supplied seawater with the elements of life.

THE BIG BOIL

Life arose on this planet during a period of crustal formation and volcanic outgassing of the atmosphere and ocean (Table 2). This was also a time of heavy meteorite bombardment. Rocky asteroids and icy comets constantly showered the early Earth, providing their unique ingredients to the boiling cauldron. The molecules of life might therefore have arrived aboard meteorites originating from elsewhere in the solar system or from deep outer space.

One of the vehicles by which organic substances traveled around the galaxy are meteoroids. They took the form of debris cast off by supernovas or the agglomeration of cosmic dust particles into small planetesimals or planetoids. Earth might have been seeded with organic compounds from cosmic debris originating from other parts of the galaxy. Intragalactic dust clouds contain organic molecules that could be incorporated into comets and meteoroids. Cosmic rays might have triggered the synthesis of organic molecules in the dark clouds even at temperatures near absolute zero.

Interplanetary space was littered with meteoroids that pounded the newborn planets. Some of this space junk might have provided organic compounds from which life could evolve. In this light, the planet might not have been the only body in the solar system to receive the seeds of life. It continues to be pelted by meteorites that contain amino acids, the precursors of proteins.

Earth contains a substantial amount of carbon, much of which originated from the interior as carbon dioxide and other carbon compounds erupting from volcanoes. Carbon was also derived from meteorites, much of which fell to Earth from primitive meteorites called carbonaceous chondrites that are essentially chunks of carbon-rich rock. They are leftovers from the formation of the solar system or are late arrivals blasted into space by supernovas. Presently, amino acids and DNA bases exist in meteorites that formed elsewhere in the solar system.

Earlier theories on the creation of life relied on the so-called primordial soup hypothesis. It states that all the precursors of life came together and, through countless combinations and permutations, an organic molecule evolved that could replicate itself. Experiments conducted in the early 1950s used spark chambers (Fig. 34) that contained the elements of the early atmosphere and ocean, and chemically combined them using an electric spark to represent lightning. These experiments produced a soup of amino acids. However, the time required for such a random event to occur naturally would require billions of years. Moreover, evidence gathered from ancient rocks on Earth, the Moon, and meteorites suggests that the amount of ammonia and methane assumed to be in the primordial atmosphere was not nearly as abundant as originally thought.

In the interior of the 4.5-billion-year-old Murchison meteorite, named for a site in Western Australia where it fell to Earth in 1969, biophysicists, who study the mechanics of life, have discovered intriguing evidence for the initiation of life. The meteorite held lipidlike organic compounds that could self-assemble into cellularlike membranes—an essential requirement for the first living cells. The meteorite is a carbonaceous chondrite believed to have broken off an asteroid formed about the same time and from similar materials as Earth. The organic chemicals provide the first unambiguous evidence of extraterrestrial amino acids. The material in the meteorite thus contains many essential components necessary for creating life. Earth is still pelted by meteorites that contain amino acids, the precursors of proteins.

The massive meteorite impacts early in Earth's history would also have made living conditions very difficult for life, which was first striving to organize proteins into living cells. The first cells might have been repeatedly exterminated, forcing life to originate again and again. Whenever primitive organic molecules attempted to arrange themselves into living matter, frequent impacts blasted them apart before they could reproduce.

Some large impactors might have generated enough heat to boil off a large portion of the ocean repeatedly. The vaporized ocean would have raised surface pressures more than 100 atmospheres. The resulting high temperatures would have sterilized the entire planet. Several thousand years elapsed before Earth cooled sufficiently for steam to condense into rain and refill the ocean basins once again, only to await the next ocean-evaporating impact. Such harsh conditions could have set back the emergence of life hundreds of millions of years.

Perhaps the only safe place for life to evolve was on the deep ocean floor. A high density of hydrothermal vents (Fig. 35) acting like geysers on the bot-

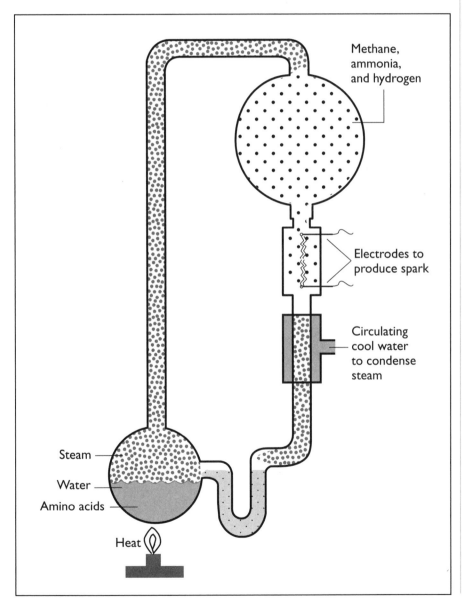

Figure 34 *The Miller-Urey experiment attempted to reproduce the early conditions on Earth when life first evolved using a spark discharge apparatus.*

Methane, ammonia, and hydrogen

Electrodes to produce spark

Circulating cool water to condense steam

Steam

Water

Amino acids

Heat

Figure 35 An active hydrothermal vent and sulfide-mineral deposits at the East Pacific Rise.

(Photo courtesy USGS)

tom of the ocean expel mineral-laden hot water heated by shallow magma chambers. These vents might have created an environment capable of generating organic reactions whereby life could have originated as early as 4.2 billion years ago. Therefore, life might not be unique to this planet alone but could exist elsewhere in the solar system as long as water and volcanic activity are present.

After discussing how this planet formed, the next chapter examines historic meteorite impact events.

3

CRATERING EVENTS
HISTORIC METEORITE IMPACTS

This chapter examines major meteorite impact events in historical geology. During Earth's long history, the planet was repeatedly bombarded by asteroids and comets. The cratering rate was much higher in the early years than today, which was fortunate for life. For if the rate of meteorite bombardment had remained high, species would have gone down much different evolutionary pathways. The meteorite bombardments would have made living conditions very difficult for early life-forms. Frequent impacts would have blasted organic molecules apart before they ever had a chance to organize into living cells.

Meteorite impacts have also been cited as the causes of extinction of species at various intervals of geologic time. Sometimes, asteroids as large as mountains or comets the size of icebergs struck Earth and inflicted so much havoc the extinction of large numbers of species was inevitable. Massive comet swarms, involving perhaps thousands of impactors striking Earth, might also explain the disappearance of species throughout geologic history.

ARCHEAN IMPACTS

The first 4 billion years, or about 90 percent of geologic time (Table 3), comprise the Precambrian era. This is the longest and least understood interval of Earth's history due mostly to major alterations of ancient rocks. The Precambrian is divided into the Hadean eon, also called the Azoic eon, meaning "time before life," from about 4.6 to 4 billion years ago; the Archean eon, meaning "time of ancient life," from about 4 to 2.5 billion years ago; and the Proterozoic eon, meaning "time of primitive life," from about 2.5 billion to 540 million years ago.

The first half-billion years of Earth's history are missing from the geologic record due to the continued destruction of the crust by heavy volcanic and meteoritic activity during the Hadean eon, when the planet was in a great turmoil. Large-scale magmatic intrusion along with numerous large meteorite impacts characterized the unusual geology of the period. About 4 billion years ago, during the height of the great meteorite bombardment, a massive asteroid landed in present-day central Ontario, centered over the Superior Province (Fig. 36). The impact was so powerful it blasted a huge crater 500 to 900 miles wide. The massive thermal energy possibly triggered the formation of the continental crust that presently makes up much of the North American continent, the oldest on Earth. Few rocks date beyond 3.8 billion years, suggesting that little continental crust formed prior to this time and was recycled into the mantle.

Small patches of continental crust might have existed on Earth's surface early in the Archean. The evidence is manifested by 4-million-year-old Acasta gneiss, a metamorphosed granite, in the Northwest Territories of Canada. This suggests the formation of the crust was well under way during this time. The continental crust contained slivers of granite that drifted freely over Earth's surface driven by vigorous tectonic plate motions.

As the restless Earth began to settle down and meteorite impacts became less pronounced, the interior gradually cooled. This allowed the formation of a permanent crust composed of a thin layer of basalt embedded with scattered blocks of granite called *rockbergs*. These slices of crust combined into stable bodies of basement rock, upon which all other rocks were deposited. The basement rocks formed the nuclei of the continents and are presently exposed in broad, low-lying, domelike structures called shields.

The Precambrian shields are extensive uplifted areas surrounded by sediment-covered bedrock called continental platforms. The platforms are broad, shallow depressions of basement complex filled with nearly flat-lying sedimentary rocks. The best-known areas are the Canadian Shield in North America (Fig. 37) and the Fennoscandian Shield in Europe. More than one-third of Australia is Precambrian shield, and large shields lie in the interiors of

TABLE 3   THE GEOLOGIC TIME SCALE

Era	Period	Epoch	Age (millions of years)	First life forms	Geology
		Holocene	0.01		
	Quaternary				
		Pleistocene	3	Humans	Ice age
Cenozoic		Pliocene	11	Mastodons	Cascades
		Neogene			
		Miocene	26	Saber-toothed tigers	Alps
	Tertiary	Oligocene	37		
		Paleogene			
		Eocene	54	Whales	
		Paleocene	65	Horses, Alligators	Rockies
	Cretaceous		135		
				Birds	Sierra
Nevada					
Mesozoic	Jurassic		210	Mammals	Atlantic
				Dinosaurs	
	Triassic		250		
	Permian		280	Reptiles	Appalachians
	Pennsylvanian		310		Ice age
				Trees	
	Carboniferous				
Paleozoic	Mississippian		345	Amphibians	Pangaea
				Insects	
	Devonian		400	Sharks	
	Silurian		435	Land plants	Laursia
	Ordovician		500	Fish	
	Cambrian		570	Sea plants	Gondwana
				Shelled animals	
			700	Invertebrates	
Proterozoic			2500	Metazoans	
			3500	Earliest life	
Archean			4000		Oldest rocks
			4600		Meteorites

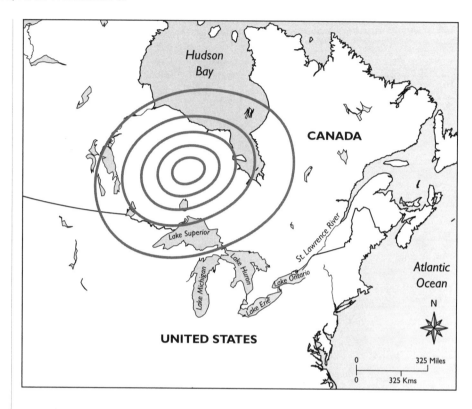

Figure 36 *The rings indicate the location of Archean impact structures in central Ontario.*

Africa, South America, and Asia as well. Many shields are fully exposed where flowing ice sheets eroded their sedimentary cover during the ice ages. Because they have been stable for billions of years, shields preserve ancient impact structures far better than do other geologic settings.

Only three sites in the world, located in Canada, Australia, and Africa, contain rocks exposed on the surface during Earth's early history that have remained essentially unchanged through geologic time. Some of the oldest rocks found on Earth in South Africa and Australia contain layers of tiny spherical silicate grains suspected of being debris from the oldest known meteorite impacts. The debris dates from around 3.5 billion years ago, during a time when large impacts played a much more significant role in shaping Earth's surface than later on. The impacts were also much more numerous and more powerful.

The extremely high temperatures generated by the impact force fused sediments into small glassy spherules. Extensive deposits of sand–sized spherules exist in the Barberton Greenstone Belt of South Africa in places more than 1 foot thick. The spherules appear to have originated from the melt created by a large meteorite impact sometime between 3.5 and 3.2 billion years ago. Spherules of a similar age also exist in the eastern Pilbara Block of Western Australia. The spherules resemble the glassy chondrules found in carbonaceous chondrites, which are primitive carbon-rich meteorites and are found in lunar soils (Fig. 38).

The widespread bed of silicate spherules is highly enriched in iridium. This isotope of platinum is extremely rare in Earth's crust but relatively abundant in asteroids and comets. In the earliest stages of Earth's formation, molten iron scrubbed out the crust's iridium and platinum along with other siderophiles (iron lovers) and carried them deep down to the core, leaving the crust virtually free of these elements. Therefore, anonymous concentrations of iridium on Earth's surface suggest it arrived aboard meteorites or comets.

Greenstone belts, a mixture of metamorphosed lava flows and sediments, occupy the ancient cores of the continents (Fig. 39). They span an area of several hundred square miles, surrounded by immense expanses of gneiss, the metamorphic equivalents of granites and the predominant Archean rock types. Their color derives from chlorite, a greenish, micalike mineral.

The existence of greenstone belts is used as evidence for early plate tectonics, the interaction of crustal plates that produces the surface features of Earth. This process might have operated in the Archean some 2.7 billion years

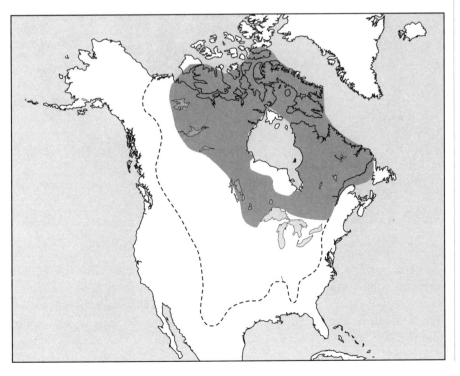

Figure 37 *The Canadian Shield* (darker area) *and platforms* (enclosed by the dashed line).

Figure 38 *A view of an astronaut's footprint in lunar soil during the* Apollo 11 *landing on the Moon on July 20, 1969.*

(Photo courtesy NASA)

ago, with small tectonic plates clashing with each other as early as 4 billion years ago. The best-known greenstone belt is the Swaziland sequence in the Barberton Mountain Land of southeastern Africa. It is more than 3 billion years old and nearly 12 miles thick. Since greenstone belts are geologically unique to the Archean, their absence after 2.5 billion years ago marks the end of the eon.

PROTEROZOIC IMPACTS

The Proterozoic was markedly different from the Archean and represented a shift to more stable geologic conditions. During the Proterozoic, Earth matured from a turbulent youth to a quiescent adulthood and acquired much of the geology that exists today. When the eon began, as much as 80 percent of the current continental crust had already formed. Continents were less erratic and welded together into a single large supercontinent called Rodinia, Russian for "motherland." Rodinia subsequently broke apart into four or five separate landmasses when the eon drew to a close.

The continents of the Proterozoic consisted of an assortment of Archean cratons, comprising pieces of ancient granitic crust. The original cratons

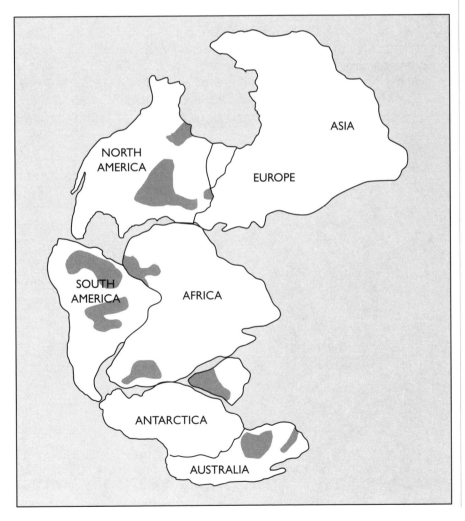

Figure 39 *Archean greenstone belts comprise the ancient cores of the continents.*

formed within the first 1.5 billion years of Earth's existence and totaled only about one-tenth of the present-day landmass. The ancestral North American continent called Laurentia assembled from several cratons (Fig. 40) beginning about 1.8 billion years ago. Most of the continent evolved in a relatively brief period of only 150 million years.

The global climate of the Proterozoic was cooler compared with the Archean. Earth experienced its first major glaciation around 2.2 billion years ago when ice practically covered the entire landmass, reaching from the poles to the tropics. Perhaps only a truly catastrophic event such as a massive meteorite impact could have thawed the planet from its frosty state. Otherwise, Earth could have remained icebound to this very day.

During this time, a huge meteorite impacted on South Africa and formed the wide Vredefort impact structure. The crater's original diameter, estimated at 100 miles or more, has been heavily modified over the last 2 billion years. The melted materials created by the impact contained a high concentration of the rare element iridium, indicating the crater must have had an extraterrestrial origin.

About 1.85 billion years ago, a large meteorite slammed into the North American continent in the present-day region of Ontario, Canada, and created enough energy to melt crustal rocks. Vast quantities of basalt and granite

Figure 40 *The cratons that make up the North American continent.*

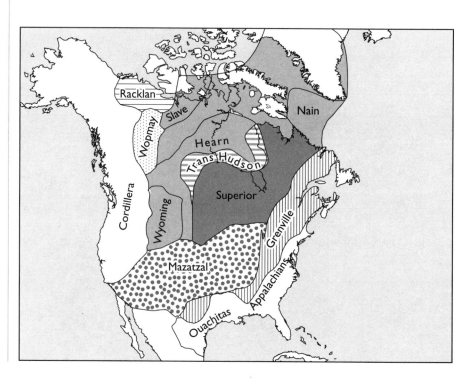

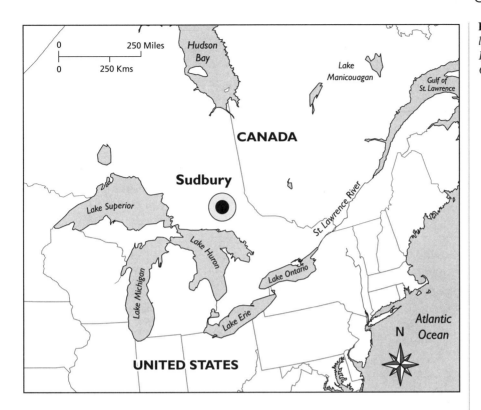

were instantly liquefied. Metals separating out of the molten rocks formed one of the world's largest and richest nickel and copper ore deposits, known as the Sudbury Igneous Complex (Fig. 41). Apparently, only an impact could have produced such an unusual abundance of ores at Sudbury. The 40-mile-long by 25-mile-wide structure comprises zones of different igneous rock types stacked like a series of elliptically shaped bowls.

The main line of evidence that Sudbury was the result of a meteorite impact were grains of shocked minerals discovered in the region along with shatter cones, distinctively striated conical rocks fractured by powerful shock waves found only at known meteorite impact sites. The location also displays one of the world's oldest astroblemes, which are ancient eroded impact structures. The original meteorite crater began with a circular shape that was later deformed by tectonic activity. It measured roughly 125 miles in diameter and is among the largest known impact structures on Earth.

PALEOZOIC IMPACTS

The Cambrian period witnessed an explosion of new species that was the most remarkable and puzzling event in the history of life. In the early Cambrian, a rapid proliferation of shelly faunas (Fig. 42) led to the progenitors of all life on Earth today. The biologic proliferation peaked about 530 million years ago, filling the ocean with a rich assortment of life. Seemingly out of nowhere and in bewildering abundance, animals appeared in astonishingly short order cloaked in a baffling array of exoskeletons. The introduction of hard skeletal parts has been called the greatest discontinuity in Earth's history and signaled a major evolutionary change by accelerating the developmental pace of new organisms. Unfortunately, many of these new species succumbed to extinction from subsequent meteorite impacts.

More than 1 mile beneath the floor of Lake Huron lies a 30-mile-wide rimmed circular structure that appears to be an impact crater blasted out by a large meteorite at least 500 million years ago. The ringed formation was first detected using magnetic sensors. It was named the Can-Am structure because it straddles the border between Canada and the United States. A crater this size would have required the impact of a 3-mile-wide meteorite with impact pressures approximating those in Earth's core. The impact structure numbers among at least 150 other known large craters gouged out during the last 500 million years.

In the early Paleozoic, all continents were dispersed around the Iapetus Sea, centered over the present-day Atlantic Ocean. From about 420 million to 380 million years ago, Laurentia collided with Baltica, the ancient European continent, and closed off the Iapetus. The collision fused the two continents

Figure 42 *Early Cambrian hard-shelled faunas.*

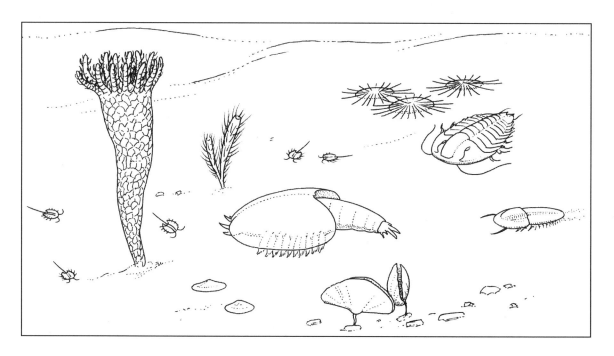

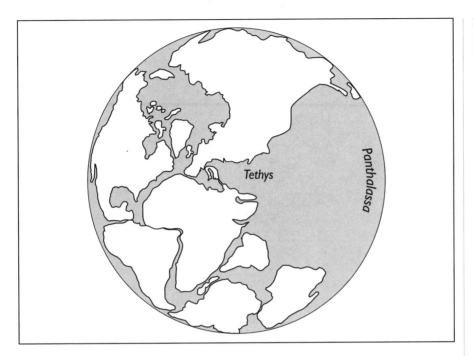

Figure 43 *The upper Paleozoic supercontinent Pangaea.*

into the great northern landmass Laurasia, named for the Laurentian province of Canada and the Eurasian continent. It included present-day North America, Greenland, Europe, and Asia.

The southern landmass called Gondwana, named for a geologic province in east-central India, included today's Africa, South America, Australia, Antarctica, and India. Gondwana linked with Laurasia to form Pangaea (Fig. 43), meaning "all lands," between 360 and 270 million years ago. The crescent-shaped supercontinent extended almost from pole to pole and was surrounded by a global ocean called the Panthalassa, meaning "universal sea." Volcanic eruptions and meteorite bombardments were frequent on Pangaea. One such asteroid 1 mile or more across slammed into Earth and dug out a crater 8 miles wide that is 70 miles south of where Chicago is today.

A major extinction event near the end of the Devonian period about 365 million years ago wiped out many tropical marine groups. The die-off of species apparently occurred over a period of 7 million years and eliminated species of corals and many other bottom-dwelling marine organisms. Primitive corals and sponges, which were prolific limestone reef builders early in the period, suffered heavily during the extinction and never fully recovered. Large numbers of brachiopod families also died out at the end of the period. While these species were vanishing, many groups such as the glass sponges and tetracorals (Fig. 44), an important group of reef-building corals, rapidly diversified.

Figure 44 *The extinct tetracorals were major reef-building corals.*

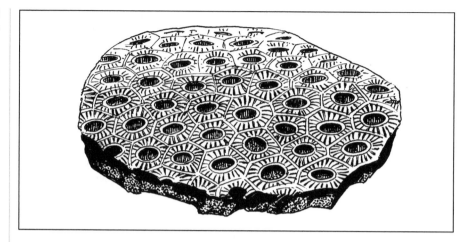

A possible prelude to the extinction was the bombardment of Earth by one or two large asteroids or comets. Evidence for the meteorite impacts is supported by the discovery of deposits containing glassy beads called microtektites in Belgium and the Hunan province of China. Microtektites form when a large meteorite impacts Earth and hurls droplets of molten rock high into the atmosphere. There they quickly cool into bits of glass that are distinct from obsidian, a natural glass created by volcanic activity. What is remarkable about these tiny, glassy beads is that they are rarely found in rocks dating much older than 40 million years.

The deposits also include an anomalous iridium content, which strongly indicates an extraterrestrial source. The Siljan crater in Sweden, approximately the same age as the microtektites, is thought to be the source of the impact deposits. The evidence suggests that meteorite bombardments appear to have contributed to many mass extinctions throughout Earth's history.

Eight large, gently sloping depressions 2 to 10 miles wide and averaging 60 miles apart are splayed across southern Illinois, Missouri, and eastern Kansas. The structures have been dated between 310 million and 330 million years old. The remarkably straight-line alignment and similarity of the pockmarks have led some geologists to conclude they formed by a series of subterranean volcanic explosions. However, no volcanic rock has been found at any of the sites.

The features are better explained as being eroded traces of a string of craters created when pieces of an asteroid or a comet broke up and slammed into Earth. A similar string of impact structures has been identified on Earth's Moon and on the moons of the outer planets. Icy comets are constantly breaking up as they pass into the inner solar system. If a comet fragmented within a few million miles of this planet, its pieces would have had sufficient time to drift apart to strike the Midwest at 60-mile intervals.

An impact would also explain many features of the sites. The rock in the depressions is folded along circular fractures, radiating from a center like a bull's-eye. Furthermore, crystals of shocked quartz were found at two sites in Missouri. These fractures form only when quartz is subjected to extreme pressures applied instantaneously. In addition, cone-shaped, overlapping splinters of rock called shatter cones that are known to form only at pressures generated by impacts were found at two other sites. The craters date from a time when large numbers of marine species mysteriously disappeared.

At the end of the Paleozoic, about 250 million years ago, perhaps the greatest extinction Earth has ever known eliminated more than 95 percent of all species, mostly marine invertebrates. Trilobites (Fig. 45), a famous group of marine crustaceans and a favorite among fossil collectors, suffered final extinction at this time. On land, more than 80 percent of the reptilian families and 75 percent of the amphibian families (Fig. 46) also disappeared. The extinctions began gradually, with a more rapid pulse at the end. Only a catastrophic event such as an asteroid impact or a huge volcanic eruption could have caused a biologic disaster on this scale.

MESOZOIC IMPACTS

At the end of the Triassic period, about 210 million years ago, a huge meteorite slammed into Earth, creating the Manicouagan impact structure in Quebec, Canada (Fig. 47). The Manicouagan River and its tributaries form a reservoir around a roughly annular structure about 60 miles across. An

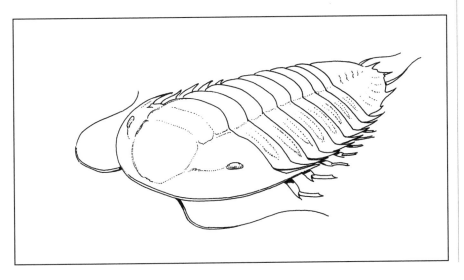

Figure 45 *Trilobites, extinct ancestors of horseshoe crabs, made prized fossils.*

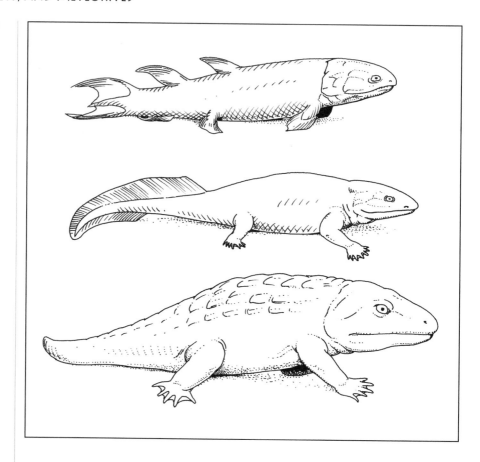

almost perfectly circular ring of water, produced when sections of the crater were flooded, surrounds the raised center of the impact structure. The site comprises Precambrian rocks reworked by shock metamorphism generated by the impact of a large celestial body. The gigantic explosion appears to have coincided with a mass extinction occurring over a period of less than 1 million years.

The Saint Martin impact structure northwest of Winnipeg, Manitoba, is 25 miles wide and mostly hidden beneath younger rocks. Three other impact structures include the 16-mile-wide Rochechouart in France, a 9-mile-wide crater in Ukraine, and a 6-mile-wide crater in North Dakota. The craters appear to have formed about the same time as the Manicouagan impact structure roughly 210 million years ago.

The date of the impacts coincides with a mass extinction at the end of the Triassic that killed off 20 percent or more of all families of animals, including nearly half the reptile families. The extinction forever changed the character of life on Earth, as the direct ancestors of modern animals emerged.

Among these were the dinosaurs (Fig. 48), which arose to dominate land life for the next 140 million years. The loss of competitors caused by the asteroid collision might have been responsible in part for the supremacy of the dinosaurs.

When Pangaea began to pull apart into the present-day continents at the beginning of the Jurassic period about 180 million years ago, the breakup of the supercontinent created three new bodies of water, the Atlantic, Arctic, and Indian Oceans. The continents were separated by a large sea called the Tethys, which established a unique circum-global circulation system that was responsible in large part for the mild climate throughout most of the period. Among the most successful of the Tethyan faunas were the prolific ammonites (Fig. 49), which were marine gastropods with a large variety of coiled shells.

Around the beginning of the Cretaceous period, roughly 130 million years ago, a large meteorite impacted the seafloor north of Scandinavia. It formed the 25-mile-wide Mjolnir crater, which lies about 1,300 feet under the Barents Sea. The identity of the crater was made by the discovery of slivers of shocked quartz and high concentrations of iridium in the sediments

Figure 47 *The Manicouagan impact structure, Quebec, Canada, from* Skylab *in 1973.*

(Photo courtesy NASA)

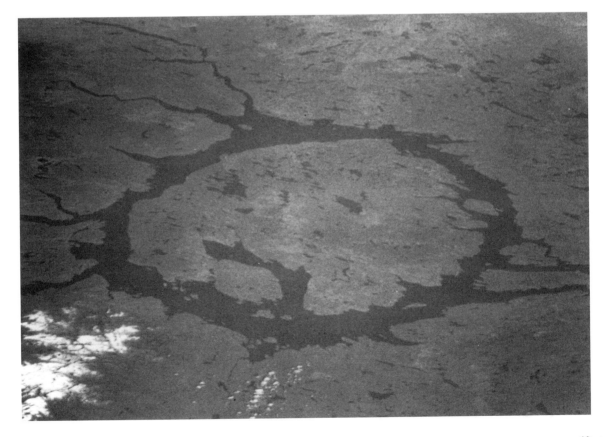

nearby. Because it impacts the seabed, the crater and surrounding debris are among the best preserved on Earth.

The world's largest impact structure appears to cover most of the western Czech Republic centered near the capital city Prague. The crater is about 200 miles in diameter and at least 100 million years old. Concentric circular elevations and depressions surround the city, suggesting the Prague basin was indeed a meteorite crater. Buildings in the city itself were constructed with rocks containing ancient impact debris. Moreover, green tektites created by the melt from an impact were found in an arc that follows the southern rim of the basin. The circular outline was discovered in a weather satellite image of Europe, and its immense size probably prevented it from being noticed by any other means.

Upheaval Dome near the confluence of the Colorado and Green Rivers in Canyon Lands National Park, Utah, appears to be a deeply eroded astrobleme,

Figure 48 *Dinosaurs dominated land life for 140 million years.*

a remnant of an ancient impact structure. It was gouged out by a large cosmic body striking Earth perhaps 100 million years ago, making the feature possibly the planet's most deeply eroded meteorite crater. Erosion has removed as much as 1 mile or more of the overlying beds since the impact.

The original crater apparently made a 4.5-mile-wide hole in the ground that has been heavily modified by deep erosion over the many years. The dome itself is a 1.5-mile-wide mound that appears to be a central rebound peak formed when the ground heaved upward by the impact. The impactor thought to measure about 1,700 feet wide crashed to Earth with a velocity of several thousand miles per hour. On impact, the meteorite created a huge fireball that would have incinerated everything within hundreds of miles.

Around 75 million years ago, an asteroid or comet landed near the present-day town of Manson in northwest Iowa and blasted out a crater about 22 miles wide, the largest known in the continental United States. At the time of the impact, the Manson area was submerged under a shallow inland sea. Today, the crater is buried beneath 100 feet of glacial till produced by successive glaciers during the Pleistocene ice ages.

When India broke away from Gondwana early in the Cretaceous period, it sped across the ancestral Indian Ocean and slammed into southern Asia around 45 million years ago. Along its journey toward Asia at the end of the Cretaceous about 65 million years ago, the subcontinent was apparently struck by a large impactor that left a 185-mile-wide circular depression on the bottom of the Indian Ocean off eastern Africa (Fig. 50).

Also during this time, a large meteorite some 6 miles across appears to have struck the north coast of Mexico's Yucatán peninsula, sending the planet

Figure 50 *The location of the crater south of the Seychelles Bank, when India drifted toward Asia 65 million years ago.*

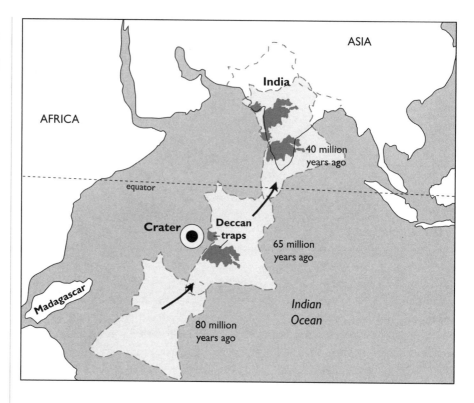

into environmental chaos. The Chicxulub structure (Fig. 51), named for the small village at its center and which means "the devil's tail" in Mayan, is one of the largest known craters on Earth, measuring between 110 and 185 miles wide. The impact crater is well hidden beneath about 1 mile of sediments. The meteorite impact left its footprints all over the world.

Sediments at the boundary between the Cretaceous and Tertiary periods (Fig. 52) on all continents contained shocked quartz grains with distinctive lamellae. They also contained common soot from global forest fires, rare amino acids known to exist only on meteorites, the mineral stishovite—a dense form of silica found only at known impact sites, and unique concentrations of iridium. The impact might have been responsible for the extinction of the dinosaurs along with more than half all other species, mostly terrestrial animals and plants.

Two or more strikes from comets knocked loose from the Oort cloud at the edge of the solar system might best demonstrate the existence of several impact craters dating to the end of the Cretaceous. One such object might have landed in the Pacific Ocean as indicated by the distinctive composition of spherules found in the sea nearby. Indeed, sediments in the Rocky Mountains dating to this time show two distinct clay layers, possibly from dual impacts. The multiple impacts might also explain why the mass extinction at the end of the Cretaceous was the most severe of the past 200 million years. Life took 5,000 years to recover from the environmental disaster.

The Hell Creek Formation in the fossil-rich badlands of eastern Montana and western North Dakota was laid down during the final 2.5 million years of the Cretaceous. During that time, the area was a swampy delta, a remnant left behind by the southern retreat of a great inland sea. A collection of terrestrial fossils from the area disputes the idea that a single catastrophic event caused the extinction of the dinosaurs and many other creatures. Furthermore, pollen fossils revealed that many plant species in the region died out rather suddenly at the end of the Cretaceous. A catalog of the number of species there along with older sites in Wyoming and Alberta, Canada, suggests that the number of dinosaur genera in the region fell from 30 to only 12 over the remaining 8 million years of the period.

Many species, including dinosaurs, appear to have already been in decline several million years prior to the end of the Cretaceous. Therefore a meteorite impact would have simply delivered the final blow. One hypothesis even suggests that a few dinosaur species might have survived the impact by

Figure 51 *The locations of possible impact structures in the Caribbean area that might have ended the Cretaceous period.*

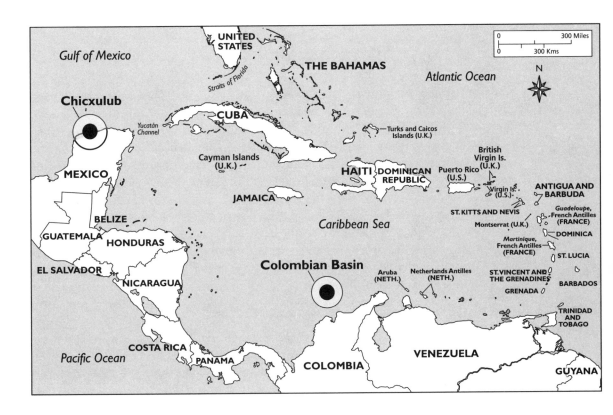

Figure 52 *The Cretaceous-Tertiary contact is at the base of the white sandstone in the center of the picture, near the foot of the hill, Jefferson County, Colorado.*

(Photo by R. W. Brown, courtesy USGS)

as much as 1 million years or more. However, no undisputed dinosaur bones have thus far been recovered in sediments above the Cretaceous, which seems to indicate that their death was indeed sudden.

CENOZOIC IMPACTS

Because water covers nearly three-quarters of Earth, most meteorites land in the ocean. Several sites have been selected as possible marine impact craters. The most pronounced undersea meteorite crater is the 35-mile-wide Montagnais structure (Fig. 53), with its distinct uplifted core, 125 miles off the southeast coast of Nova Scotia. The circular formation was first explored by oil companies in the 1970s. Instead of petroleum, they found rocks melted by a sudden shock of a large meteorite impact. Its melt rocks are also chemically similar to the tektite glass layer off the coast of New Jersey.

The crater is 50 million years old and closely resembles craters on dry land, only its rim is 375 feet beneath the sea and the bottom of the crater plunges to a depth of 9,000 feet. The crater was created by a large meteorite as wide as 2 miles. The impact raised a central peak similar to those in the interiors of craters on the Moon. The structure also contained rocks that had been melted by a sudden shock. Such an impact would have sent a tremendous tsunami crashing down on nearby shores. Because of its size and location, the crater was thought to be a likely candidate for the source of the North American tektites. However, its age estimates appear to be several million years too young to have created the tektites.

Around 37 million years ago, two or three large meteorite impacts might have caused the extinction of the archaic mammals. These were large, grotesque-looking animals that were apparently overspecialized and could not adapt to the changing climatic conditions possibly triggered by the impacts. Concentrations of microtektites and anomalously large amounts of iridium, indicative of meteorite impacts, were found in sediments dating near the end of the Eocene epoch. A possible site for one of the impacts is a 15-mile by 9-mile crater located approximately 80 miles east of Atlantic City, New Jersey. The meteorite impact sent a giant wave crashing over coastal areas from New Jersey to North Carolina.

Another meteorite apparently splashed down in the Atlantic Ocean off the Virginia coast, producing a huge wave perhaps more than 100 feet high. The tsunami apparently gouged out of the seafloor an area as large as Connecticut and deposited a 200-foot-thick layer of boulders, some measuring 3 feet wide, that are now buried under 1,200 feet of sediment. Within this boulder layer are glassy rocks or tektites and mineral grains bearing shock features, suggesting a meteorite crashed into the submerged continental shelf.

Perhaps the largest extraterrestrial impact crater in North America lies hidden below the Chesapeake Bay. The 50-mile-wide structure, the seventh

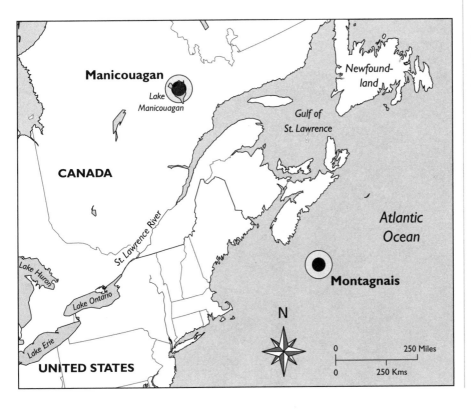

Figure 53 *The location of the Manicouagan and Montagnais impact structures in North America.*

largest so far identified on Earth, formed about 35 million years ago apparently by a meteorite up to 2.5 miles across. The tremendous impact churned up rock layers and spewed out small, glassy fragments (tektites) that were scattered over much of the eastern seaboard and thrown as far away as South America. When the meteorite hit, it left a huge depression in the continental shelf, which filled with river-borne sediment when sea levels fell. The Chesapeake Bay formed millions of years later when the ocean rose again.

About 23 million years ago, an asteroid or comet slammed into Devon Island in the Canadian Arctic with such a force that rock from more than half a mile underground shot skyward. The meteorite gouged out a 15-mile-wide hole called the Haughton Crater, and pulverized granite gneiss fell back to the ground as hot breccia. Plant and animal life ceased to exist within a radius of perhaps 100 miles. At that time, the area was much warmer and lusher with spruce and pine forests. Today, the crater is used as a test bed for future excursions to Mars, which has strikingly similar craters (Fig. 54).

Figure 54 *Martian craters viewed by the* Viking 1 *orbiter.*

(Photo courtesy NASA)

A contemporary meteorite impact might have been responsible for creating the Everglades on the southern tip of Florida (Fig. 55). The region contains a swamp and forested area surrounded by an oval-shaped system of ridges

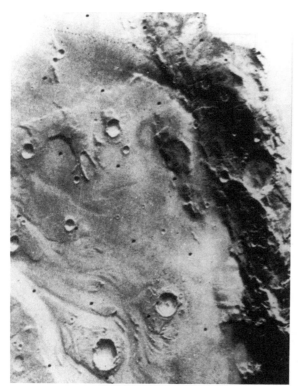
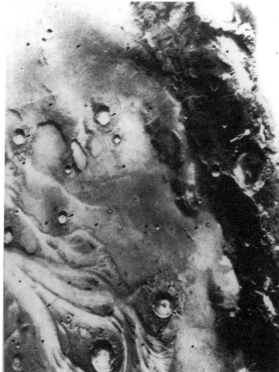

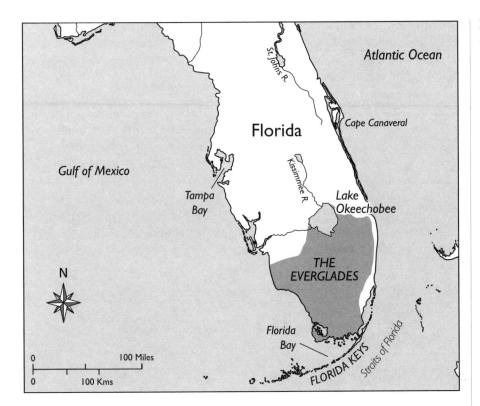

upon which rest most of southern Florida's cities. A layer of limestone measuring upward of 1,000 feet thick found in the surrounding areas is suspiciously missing over most of the southern part of the Everglades. A giant, oval-shaped coral reef, dating around 6 million years old, is buried beneath the rim surrounding the Everglades. The reef probably formed around the circular basin gouged out by a meteorite impact. A large meteorite appears to have slammed into the limestone and fractured the rocks, which were submerged under 600 feet of water. The impact would also have generated an enormous tsunami that swept debris far out to sea.

About 3.3 million years ago, a cosmic body perhaps a half mile in diameter hit the ocean just off the central coast of Argentina, producing a now-buried crater possibly 12 miles wide. The impact fused loess (a windblown deposit) into glassy slabs and flung them across at least 35 miles of coastline. The glass has streaky flow patterns typical of rapidly cooled impact glass. The impact occurred near the time of an abrupt, temporary cooling of the Atlantic and Pacific Oceans. Furthermore, a major sudden extinction at about the time of the impact wiped out 36 genera of mammals.

A major asteroid apparently landed in the Pacific Ocean roughly 700 miles westward of the tip of South America about 2.3 million years ago.

Although no crater has been found, an excess of iridium in sand–sized bits of glassy rock discovered in the region suggests an extraterrestrial origin. The impact blasted away at least 300 million tons of debris, corresponding to an object 1,800 feet or more in diameter.

The explosion from the impact would have been 100 times more powerful than the largest hydrogen bomb, with devastating consequences for the local ecology. Moreover, geologic evidence indicates Earth's climate changed dramatically between 2.2 and 2.5 million years ago, when glaciers began to roam across large portions of the Northern Hemisphere. Perhaps the impact was responsible in part for triggering the Pleistocene ice ages.

Meteor Crater (Fig. 56), also called Barringer Crater, lies near present-day Winslow in northern Arizona. It formed about 50,000 years ago when a large meteorite measuring around 150 feet wide and weighing some 300,000 tons impacted at a velocity of about 30,000 miles per hour and vaporized the

Figure 56 *Meteor Crater, Coconino County, Arizona.*

(Photo courtesy USGS)

Figure 57   *The Wolf Creek meteorite from Western Australia showing crack development on the cut surface.*

(Photo by G. T. Faust, courtesy USGS)

desert sands. The meteorite impact ejected nearly 200 million tons of rock, forming a large mushroom-shaped cloud of debris, and excavated a crater measuring about 4,000 feet across and 600 feet deep.

A steep crater rim formed by upraised sedimentary layers rises 150 feet above the desert floor. Pulverized rock blanketed the area around the crater to a depth of 75 feet. Because it lies in the desert, Meteor Crater is the best-preserved impact structure of its size in the world. It has escaped erosion that has destroyed all but the faintest signs of most other impact structures.

The New Quebec Crater in Quebec, Canada, is the world's largest known meteorite crater where actual meteorite debris has been found. It has a diameter of about 11,000 feet and a depth of about 1,325 feet, over twice the size of Meteor Crater. The crater contains a deep lake, whose surface is 500 feet below the crater rim. The impact structure is estimated at only a few thousand years old. It is well preserved because little changes in the frigid tundra.

Another relatively young impact structure called the Wolf Creek Crater is located on the northern edge of the Great Sandy Desert in Western Australia south of Halls Creek. It was first discovered from aircraft and originally thought to be of volcanic origin. The crater has a diameter of 2,800 feet and a depth of 140 feet. Several large pieces of the meteorite (Fig. 57), some weighing more than 300 pounds, were discovered near the crater. Because the crater is located in the desert, it is well preserved. The desert has preserved another young crater that measures more than 1 mile across and 220 feet deep lying near Talemzane in the Sahara.

A meteorite appears to have blasted a 1-mile-wide hole in the middle of Nebraska as little as 3,000 years ago. The 80-foot-deep circular depression located on a farm 12 miles west of Broken Bow represents the weathered remains of an impact crater, most of which lies beneath the soil. Weathering has eroded and filled in the original crater, thought to be as much as 300 feet deep. Buried fragments of glass posited to be fallout material melted and ejected by the impact were found more than 1 mile from the crater rim. While an impact of this magnitude would not have significantly altered the climate worldwide, the explosion itself would have terrified the Native Americans living nearby.

After discussing meteorite impacts in Earth's history, the next chapter takes a look at meteorite craters on the other planets.

4

PLANETARY IMPACTS
EXPLORING METEORITE CRATERS

This chapter examines meteorite craters on other planets and moons of the solar system. Finding similarities among the nine known planets and 60 some satellites is often difficult. About the only real commonality they have are a profusion of meteorite impacts. Among the most striking features of the inner rocky planets as well as the moons of the outer planets are their enormous impact craters hundreds of miles wide and several miles deep, greater than any found on Earth.

Giant impacts early in the solar system's history could have sent the planets down vastly different evolutionary paths, resulting in a wide variety of orbital motions. The many pockmarks preserved on the faces of Earth's moon, Mars, Mercury, and the moons of the outer planets demonstrate that the impacts were most frequent in the early days of the solar system. By studying the other bodies circling the Sun, an accurate account of the impact history of Earth can be established.

LUNAR CRATERS

Earth and its moon together represent a twin planetary system formed some 4.6 billion years ago. Early in its history, the Moon's surface was melted by a

massive meteorite bombardment responsible for most of its terrain features
(Fig. 58). However, because the impact craters tend to overlap, any regular, recognizable pattern is probably lost, leaving doubts about the accuracy of their
ages. The numerous rayed craters on the Moon's far side appear to be relatively
young because the bright rays of material emanating from them would have
darkened during the last billion years. These findings suggest that a much
larger-than-expected population of asteroids up to half a mile wide or more
come within the vicinity of Earth and its moon.

   Ancient large impacts produced craters as much as 250 miles across,
destroying most of the Moon's original crustal rocks. The Moon has 35 impact
basins larger than 185 miles wide, about the size of the biggest known crater
on Earth. A mammoth depression on the Moon's South Pole called the Aitken
basin is 7.5 miles deep and 1,500 miles across, or about one-quarter of the
Moon's circumference. It appears to have been initially created by a gigantic

asteroid or comet that penetrated deep into the mantle. The *Clementine* space-craft in orbit around the Moon detected what appeared to be water ice thought to be up to several hundred feet thick inside the massive crater. This might someday serve as a source of water for future lunar bases.

Beginning about 4.2 billion years ago and continuing for several hundred million years, huge basaltic lava flows welled up through the weakened crater floors and flooded great stretches of the Moon's surface, filling and sub-merging many meteorite craters. The basalt lava flows, covering some 17 percent of the lunar surface, hardened into smooth plains called maria that, when viewed from Earth, indeed resemble seas. The composition of the basalts indicates they originated from a deep-seated source. Massive outpourings of molten rock onto the surface produced oceans of lava.

Meteorite craters at the edges of maria have experienced recent land-slides, with material from the walls or rims slumping toward the center of the cavities. Dust kicked upward or volatile gases escaping from the Moon's sur-face in the aftermath of a landslide might account for many mysterious sight-ings. These strange phenomena include bright flashes, red and blue glows, and patches of mist or fog emanating from certain sites on the Moon reported as far back as the Middle Ages.

Narrow sinuous depressions in the lava flows called rilles emanate from impact craters (Fig. 59). In some areas, wrinkles break the surface of the lava flows, possibly caused by moonquakes, which explains the extensive seismic activity sensed by lunar probes. Areas of previous volcanic mountain building exist throughout the Moon's surface, with ridges reaching several hundred feet high and extending hundreds of miles. The last of the basalt lava flows hard-ened some 3 billion years ago. Except for numerous fresh meteorite impacts, the Moon looks much the same today as it did then.

Rocks on the surface of the Moon (Fig. 60) vary in age from 4.5 to 3.2 billion years old. The oldest rocks are primitive and have not changed signifi-cantly since they first originated from molten magma. The so-called Genesis Rock formed the original lunar crust and comprised a coarse-grained feldspar granite that originated from magma deep within the Moon's interior. The youngest rocks are volcanic in origin and were melted and reconstructed by giant meteorite impacts. When the basalt flows ended, new lunar rock forma-tion appears to have ceased. No known rocks are younger than 3.2 billion years.

The lunar rocks are igneous, meaning they derived from molten magma, and form the regolith composed of loose rock material on the surface. The regolith is generally about 10 feet thick but is thought to be thicker in the lunar highlands, which were especially battered during the great meteorite bombardment. The rocks include coarse-grained gabbro basalt, meteoritic impact breccia, pyroxene peridotite, glass beads known as chondrules derived from meteorite impacts, and dust-sized soil material from meteorite-blasted

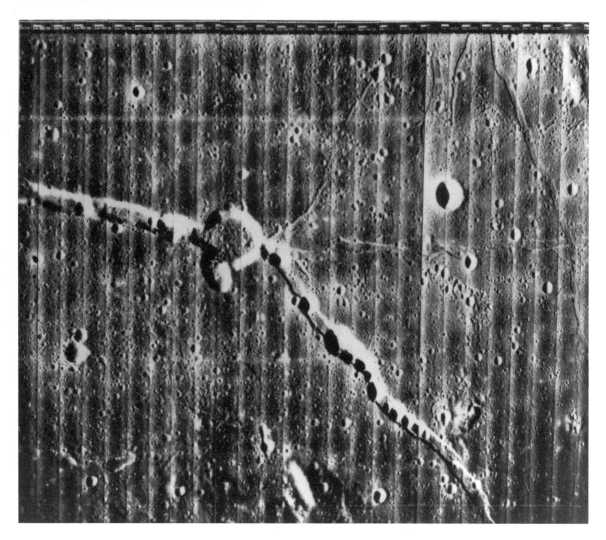

Figure 59   *The lunar crater Hyginus and Hyginus Rile.*

(Photo by D. H. Scott, courtesy USGS and NASA)

rocks. Because of its dark basalts, the Moon's surface is a poor reflector of sunlight, with an albedo of only about 7 percent. This feature makes the Moon one of the darkest bodies in the solar system.

MERCURIAN CRATERS

Mercury and the Moon are strikingly similar in appearance (Fig. 61). This observation prompted speculation that the planet might have once been the moon of Venus, which also resembles Earth in many of its attributes. Like the Moon, Mercury is heavily scarred by meteorite craters from a massive bombardment early in the creation of the solar system. Images of Mercury could easily be mistaken for the far side of the Moon. However, Mercury lacks the jumbled mountainous regions and wide lava plains, or maria, on the Moon. Instead, it has long, low, winding cliffs resembling fault lines several hundred miles long.

Mercury is heavily scarred with meteorite craters from a massive bombardment some 4 billion years ago. The huge meteorites broke through the fragile crust and released torrents of lava that paved over the surface. Several large craters are surrounded by multiple concentric rings of hills and valleys, possibly originating when a meteorite impact caused shock waves to ripple outward, similar to waves formed when a stone is tossed into a quiet pond. The largest of these craters, named Caloris, is 800 miles in diameter and 3.6 billion years old. Objects probably struck Mercury with higher velocities than those hitting the other planets due to the Sun's stronger gravitational attraction at this close range.

Figure 60 *A large lunar boulder field at Taurus-Littrow from* Apollo 17 *in December 1972.*

(Photo courtesy NASA)

Figure 61 *The heavily cratered terrain on Mercury from* Mariner 10 *in March 1974.*

(Photo courtesy NASA)

Mercury and the Moon share very similar compositions as well, comparable to the interior of Earth. Mercury has an appreciable magnetic field, indicating the presence of a large metallic core, which accounts for the planet's high density. The core is twice as massive relative to Mercury's size as that of any other rocky planet. Above the core lies a relatively thin silicate mantle, comprising only one-quarter of the planet's radius. Mercury could have formed an iron core of conventional proportions while much of its outer rocky mantle was blasted away by large planetesimal impacts.

Unlike Earth and Mars, which might have acquired high rotation rates by glancing blows from large impactors early in their formation, Mercury rotates on its axis once every 59 Earth-days and revolves around the Sun every 88 Earth-days. This unusual orbital feature requires Mercury to orbit twice around the Sun before completing a single day.

Mercury also experiences the widest temperature extremes of any planet. During the day, temperatures soar to 300 degrees Celsius. At night, they plummet to −150 degrees. The large temperature difference is due to the planet's proximity to the Sun, a highly elliptical orbit that swings from 29 million to 43 million miles from the Sun, a slow rotation rate of 1.5 times for every revolution—thereby keeping its dark side away from the Sun for long periods, and a lack of an appreciable atmosphere to diffuse the heat around the planet.

Mercury lacks a significant atmosphere because gases and water vapor formed by volcanic outgassing or by cometary degassing are quickly boiled off due to the excessive heat and low escape velocity. The tenuous atmosphere that remains contains small amounts of hydrogen, helium, and oxygen, probably originating from the direct infall of cometary material and outgassing of the few remaining volatiles in the interior. Traces of water ice might still exist in permanently shaded regions near the poles. Because of its small size, which allowed all the planet's internal heat to escape into space early in its life, Mercury is now a tectonically dead planet.

VENUSIAN CRATERS

In many respects, Venus and Earth are sister planets. Venus is nearly the same size and mass as Earth, only it contains a thick atmosphere composed almost entirely of carbon dioxide. The air pressure at the surface is nearly 100 times that on Earth. Because of its high density, the gas is more like an ocean than an atmosphere. The thick atmosphere shields the planet's surface from impactors less than half a mile in diameter, which would otherwise gouge out craters up to 9 miles wide.

The surface of Venus is relatively young, with an average age of less than 1.5 billion years. This is comparable to Earth some 3 billion years ago before plate tectonics began to change its identity. The jagged surface on Venus (Fig. 62) appears to have been shaped by deep-seated tectonic forces and volcanic activity eons ago, providing a landscape totally alien to that of Earth. Venus was thought to have counterparts to the rifts and collision zones created by active plate tectonics as on Earth. However, more detailed images from *Magellan* suggest that the Venusian surface is a single shell largely devoid of global plate tectonics. Rifting causes a small percentage of the heat flow from Venus. In contrast, on Earth, 70 percent of the interior heat is lost by seafloor spreading. Venus therefore appears to be a dry, fiery planet whose surface is locked in an immobile shell and unable to shift its crust the way Earth does.

Mountains built by crumpled and fractured crust bear a remarkable resemblance to the Appalachians, formed by the collision of North America and Africa. The 36,000-foot-high Maxwell Montes dwarfs Mount Everest by more than 1 mile. Faults resembling California's San Andreas shoot through the surface, displacing large chunks of crust. The Northern Hemisphere is smooth, only lightly cratered, and dotted with numerous extinct volcanoes. Although Venus possesses continental highlands in the Northern Hemisphere, its ocean basins lack one essential ingredient—water.

Large volcanic structures on the surface of Venus suggest its volcanoes are much more massive than those on Earth. To hold up these gigantic volcanoes,

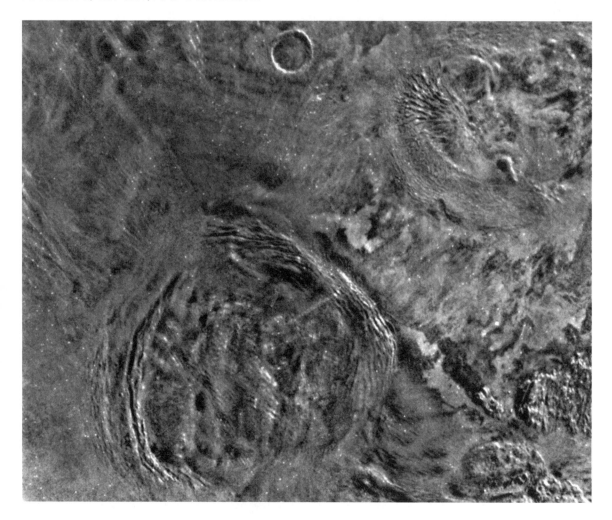

Figure 62 *Radar image of Venus's northern latitudes by Vernera 15.*

(Photo courtesy NASA)

the lithosphere beneath them would have to be 20 to 40 miles thick. Great circular features are as wide as several hundred miles and relatively low in elevation. However, instead of being large meteorite craters, these structures are attributed to collapsed volcanic domes. This results in huge, gaping holes called calderas surrounded by folds of crust as though immense bubbles of magma had burst through to the surface.

Enormous elevated, plateaulike tracts rise 3 to 6 miles above the surrounding terrain. The Beta Regio region (Fig. 63) appears to contain many large volcanoes, some of which are up to 3 miles high. A broad shield volcano known as Theia Mons is more than 400 miles wide, greater than any volcano on Earth. Large anomalies in the gravity field seem to indicate that rising plumes of mantle rock that feed active volcanoes buoy up some highlands on Venus. A great rift valley measuring 8 miles deep, 175 miles wide, and 900 miles long could well be the grandest canyon in the solar system (Fig. 64). Great floods that raged across the surface eons ago might have carved it out.

Elongated ridges and circular depressions more than 50 miles in diameter might have resulted from large meteorite impacts. Venus has some 1,000 craters spread randomly over its surface. Impacts from the first 3.7 billion years of the planet's history have been eradicated. Most of the Venusian craters appear to be fresh. Volcanism, folding, and faulting have disrupted about 20 percent of them. Most of the impact craters on Venus appear to have been erased about 800 million years ago, when widespread volcanic eruptions paved over a large portion of the planet and created vast volcanic plains.

The surface on Venus appears to be remarkably flat, much more so than that of Earth or Mars. Two-thirds of the planet has a relief of less than 3,000 feet. The density of surface rocks is identical to terrestrial granites. The composition of the soil is similar to basalts on Earth and its moon. Scattered rocks on the surface are angular in some places and flat and rounded at other locations, implying strong wind erosion. On the rugged equatorial highlands of Aphrodite Terra is the jumbled debris of a massive landslide that matches the largest avalanches on Earth.

Figure 63   *The central part of Beta Regio, from the Venus probe* Magellan.

(Photo courtesy NASA)

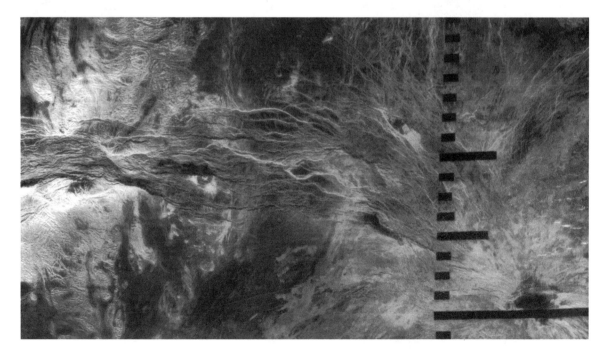

Figure 64   An artist's rendition of the Venus rift valley, which at 3 miles deep, 175 miles wide, and 900 miles long is the largest canyon in the solar system.

(Photo courtesy NASA)

MARTIAN CRATERS

Mars is divided into two very distinct geographies. The southern hemisphere is rough, heavily cratered, and traversed by huge channel-like depressions where massive floodwaters apparently flowed some 700 million years ago. The southern highlands is a highly cratered region that resembles the Moon but lacks the scars that mark the planet's northern face. The northern hemisphere is smooth, only lightly cratered, and dotted with numerous extinct volcanoes.

Nanedi Vallis, a deeply cut, sinuous channel in the Martian crust, is perhaps the strongest evidence that water existed on the planet's surface for prolonged periods. The giant Argyle Planitia impact basin in the southern highlands is about 750 miles wide and more than 1 mile deep. It shows evidence that liquid water once flowed on Mars. It contains layers of material that appear to be sediment from a huge body of water held by the basin millions of years ago. Three networks of channels lead into the basin from the south, whereas other channels slope northward out of the basin.

As on Earth, Mars has two polar ice caps, which contain about 300,000 cubic miles of water ice. The southern ice cap might have once grown large enough for water to flow into the basin, forming a huge, icy lake that spilled over the northern side of the basin and carved out deep channels. The northern ice cap holds a significant amount of the planet's water. If melted, it would cover the entire planet to a depth of about 40 feet. At one time, Mars had enough water to have created a global ocean up to 3,000 feet deep.

A several-miles-thick mixture of frozen carbon dioxide, water ice, and windblown dust caps the polar regions. Branching tributaries resembling dry riverbeds span the Martian surface. Apparently, heat generated by geothermal activity or meteorite bombardment melted subterranean ice, producing great floods of water and flowing mud that carved out colossal ditches rivaling those on Earth. The largest canyon, Valles Marineris (Fig. 65), measures 3,000 miles long, 100 miles wide, and 4 miles deep and could hide several Grand Canyons. The ravine is thought to have formed by slippage of the crust along giant faults accompanied by volcanism.

Mars has two tiny moons, Deimos and Phobos (Fig. 66), that are an enigma. The Martian satellites are oblong chunks of rock in nearly circular orbits. Their small sizes, blocky shapes, and low densities suggest they were captured bodies from the nearby main asteroid belt. However, such a capture seems highly remote. For this to happen twice would seem to be nearly impossible. Instead, the moons might be the last surviving remnants of a ring of debris blasted into orbit when a huge asteroid more than 1,000 miles wide collided with Mars some 4 billion years ago. Phobos, which is not much more than an irregular rock 14 miles across, is also in a decaying orbit and will fall to Mars in about 30 million years.

The Mars *Surveyor* spacecraft showed that much of the planet's northern hemisphere is a low-lying plain thousands of miles wide and roughly centered on the north pole, while the rest of the planet is ancient highlands. Perhaps a massive body slammed into the northern hemisphere and altered the surface, which would also explain the smaller number of impact craters. Another idea is that Earthlike tectonic forces and perhaps even an ancient ocean have shaped the northern lowlands. The relief of the land is remarkably low, making the region the flattest surface in the solar system.

A large number of volcanoes dot the Martian northern hemisphere. The largest is Olympus Mons (Fig. 67 and Table 4), whose broad base could cover the state of Ohio. It rises 75,000 feet, more than twice as high as Mauna Kea, Hawaii, the tallest volcano on Earth. The Martian volcanoes closely resemble the shield volcanoes that built the main island of Hawaii. Their extreme size apparently results from the absence of plate movements.

Rather than form a chain of relatively small volcanoes as though assembled on a conveyor belt when a plate moved across a volcanic hot spot, a sin-

Figure 65 *A mosaic of the Mars surface at the west end of the Valles Mariners canyon system. These two canyons are over 30 miles wide and nearly 1 mile deep.*

(Photo courtesy NASA)

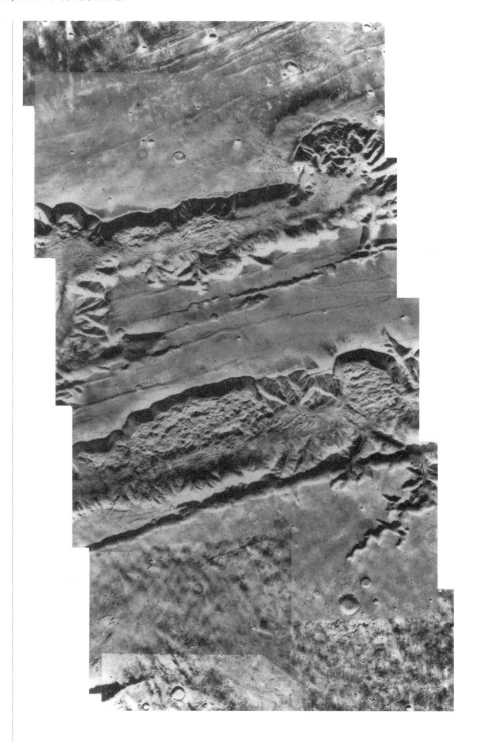

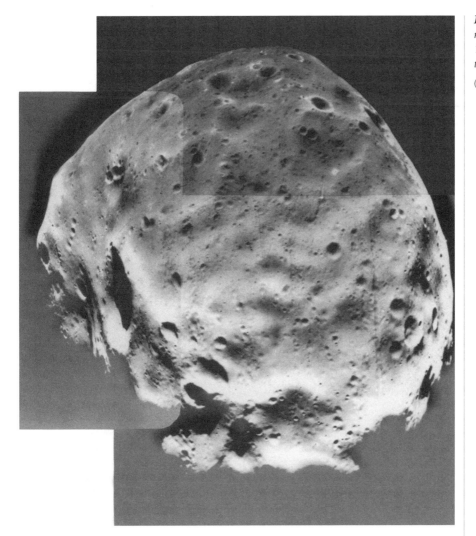

Figure 66   *Mars's inner moon Phobos, measuring 13 miles across, is thought to be a captured asteroid.*

(Photo courtesy NASA)

gle very large cone developed as the crust remained stationary over a magma body for long periods. Apparently, no horizontal crustal movement of individual lithospheric plates occurs at present. The crust is probably much colder and more rigid than that on Earth, which explains the lack of folded mountain ranges.

Meteorite impact craters are notably less abundant in regions where volcanoes are most numerous. This indicates that much of the volcanic topography on Mars formed after the great meteorite bombardment some 4 billion years ago. However, for Mars to exhibit recent volcanism, pockets of heat would have to lie just below the surface. The southern hemisphere has a more highly cratered surface comparable in age to the lunar highlands, which are

about 3.5 to 4.0 billion years old. The discovery of several highly cratered and weathered volcanoes indicates that volcanic activity began early and had a long history. Even the fresh appearing volcanoes and lava planes of the northern hemisphere are probably very ancient.

Although Mars is in proximity to the asteroid belt and the number of meteorite impacts were expected to be high, organic matter from infalling carbonaceous chondrites should have been abundant. Yet not a single trace of organic compounds was uncovered by Mars landers. Apparently, because the thin Martian atmosphere has only about 1 percent of the atmospheric pressure of Earth, organic compounds are actively destroyed by the Sun's strong ultraviolet radiation.

TABLE 4   MAJOR VOLCANOES OF MARS

Volcano	Height (miles)	Width (miles)	Age (million years)
Olympus Mons	16	300	200
Ascraeus Mons	12	250	400
Pavonis Mons	12	250	400
Arsia Mons	12	250	800
Elysium Mons	9	150	1,000–2,000
Hecates Tholus	4.5	125	1,000–2,000
Alba Patera	4	1,000	1,000–2,000
Apollinaris Patera	2.5	125	2,000–3,500
Hadriaca Patera	1	400	3,500–4,000
Amphitrites Patera	1	400	3,500–4,000

Figure 68   *The Mars* Pathfinder *landing site in Ares Vallis showing evidence of wind erosion and sedimentation.*

(Photo courtesy NASA)

The entire planet displays evidence of wind erosion and sedimentation. Violent seasonal dust storms responsible for the planet's red glow whip up winds of 170 miles per hour for weeks at a time. Winds play an important role by stirring the surface sediments and scouring out ridges and grooves in the surface. Deep layers of windblown sediment have accumulated in the polar regions, and dune fields larger than any on Earth lie in areas surrounding the north pole.

Most of the Martian landscape is a dry, desolate wasteland etched by deep valleys and crisscrossing channels, bearing the unmistakable imprint of flowing water sometime in the planet's long history. Channels dug out of the surface seem to imply that the climate in the past was significantly different from that of today. Flowing streams might have fanned out of the highlands and disappeared into the deserts as do flash floods on Earth.

The site of the Mars *Pathfinder* probe sits at the mouth of an ancient flood channel 1 mile deep and 60 miles wide called Ares Vallis (Fig. 68). Apparently, the equivalent of all the water in the Great Lakes once poured through the valley in only a few weeks, draining a large section of the Martian surface in a catastrophic flood. The great flood transported a variety of interesting rocks and other material from a long distance away. Rounded pebbles, cobbles, and conglomerates tumbled by running water are a sure sign that floods rushed through the region.

OUTER PLANETARY CRATERS

The outer large bodies of the solar system are referred to as giant gaseous planets due to their thick atmospheres. The only exception is Pluto, considered an asteroid or a moon of Neptune knocked out of orbit by a collision with a comet. Because of their low densities, gases are believed to make up the bulk of the planets' mass, indicating their evolution was significantly different from the inner, dense, terrestrial planets. This is especially true for Jupiter (Table 5). Moon-sized building blocks comprising rock and ice made up the cores, which grew massive enough to gather gases to build up the outer layers. Uranus is the strangest of all and seems to have taken a heavy, off-center hit by a massive planetesimal falling into it, tilting the planet over onto its side.

Although the large planets are spectacular in their own right, they are not as geologically significant as their moons. Most of the moons of Jupiter, Saturn, and Uranus appear as though they were resurfaced, cracked, modified by flows of solid ice, and heavily bombarded by meteorites.

Jupiter and its 15 moons resemble a miniature solar system. The four largest moons first discovered by Galileo in 1610 and thus named the Galilean moons travel in nearly circular orbits with periods ranging from two to 17 days. The largest of the Galilean moons, Callisto and Ganymede, are about the size

TABLE 5   CHARACTERISTICS OF JUPITER'S ATMOSPHERE

Characteristic	Belt	Zone	Great Red Spot
Infrared energy	Hot	Cold	Cold
Cloud height	Low	High	High
Vortex	Cyclonic	Anticylonic	Anticylonic
Pressure	Low	High	High
Temperature	Cold	Hot	Hot
Vertical winds	Down	Up	Up
Cloud types	Low, thin	High, thick	High, thick
Cloud color	Dark	Light	Orange

of Mercury. The two smaller moons, Europa and Io, are about the size of Earth's moon. The most impressive feature of Callisto, the outermost of the Galilean satellites, is that its surface is almost completely saturated with craters, resembling the far side of Earth's moon. A prominent bull's-eye structure (Fig. 69) appears to be a large impact basin on a frozen crust composed of dirty ice.

The surface of Ganymede, the largest moon in the solar system, is less cratered than Callisto and resembles the near side of Earth's moon, with densely cratered regions and smooth areas where young, icy lavas cover the scars of older craters. Some smooth areas are heavily cratered and splintered by fractures. Numerous grooves, faults, and fractures suggest tectonic activity in the not too distant past. A long band in the Arbela Sulcus region suggests that sections of crust slid past each other, as if they had glided over a layer of warm, pliable ice. A complex intersecting network of branching bright bands crisscross the moon's surface with meteorite craters in between (Fig. 70).

Ganymede is the only moon known to have a significant magnetic field. Like Earth's moon, the satellite keeps the same side facing Jupiter at all times. On the side facing away from the planet is a large circular region of dark, cratered terrain with closely spaced parallel ridges and troughs from 3 to 10 miles across. They formed when a huge meteorite smashed into the soft crust and fractured it, producing an enormous pattern of concentric rings. Later, water filled the fractures and froze, becoming one of the most impressive sights the solar system has to offer.

An intricate tangle of crisscrossing ridges on Europa, the second Galilean satellite outward from Jupiter and about the same size as Earth's moon, suggests icy volcanic eruptions resembling midocean ridges on Earth. Ice rises to the surface at the central ridge and spreads away to form new crust. Europa appears to have an icy crust broken into giant ice floes at least 6 miles thick. Below the crust might be liquid water heated by undersea volcanoes, the only other known water ocean in the entire solar system.

The surface of Europa is crisscrossed by dark stripes and bands thousands of miles long and up to 100 miles wide (Fig. 71) that appear to be fractures in the icy crust filled with material erupted from below. The surface is remarkably free of impact craters, indicating the moon's surface gets paved over every 10 million years as new material rises from below and buries other features including meteorite craters. The moon probably formed after the great meteorite bombardment period, some 4 billion years ago, when impacts were much more prevalent than today.

Images from the *Galileo* spacecraft reveal a complex network of ridges and fractures, some of which resemble features formed by plate tectonics on

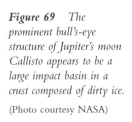

Figure 69 *The prominent bull's-eye structure of Jupiter's moon Callisto appears to be a large impact basin in a crust composed of dirty ice.*

(Photo courtesy NASA)

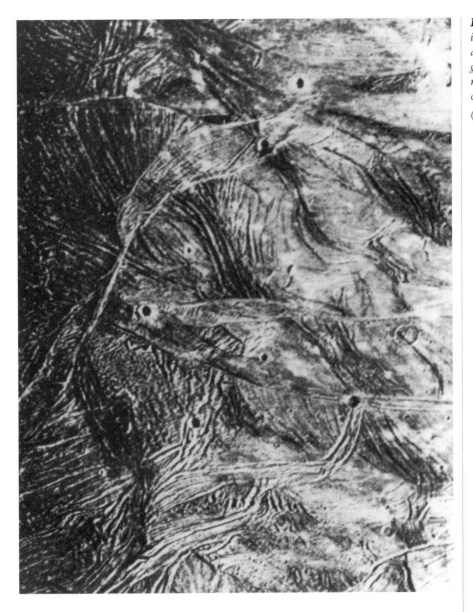

Figure 70 *Ganymede is Jupiter's largest moon and shows ridges and grooves that probably resulted in deformation of the icy crust.*

(Photo courtesy NASA)

Earth. In places, the surface of Europa looks like a giant jigsaw puzzle that has been pulled apart. Through the cracks flowed huge outpourings of slushy ice that buried all Europa's terrain features, including its impact craters. Unlike the bowl-shaped impact craters on other bodies, the largest impact features on Europa have a central smooth patch surrounded by concentric rings. The impactors apparently penetrated the rigid outer ice crust. Melted water and slush quickly filled in the craters while the impactors fractured the surface into rings that are like a frozen record of a rock tossed into a quiet pond.

Io, the innermost satellite, is the most intriguing of the Jovian moons. Its size, mass, and density are nearly identical to that of Earth's moon. Widespread volcanism over the entire moon's surface produces more than 100 volcanoes (Fig. 72), making Io possibly the most volcanically active body in the solar system. The surface of Io is remarkably young. The moon is the only one in the solar system with essentially no impact craters. The nearly total lack of craters indicates the surface of Io has been paved over with large amounts of lava within the last million years. Several major volcanoes erupt at any given moment, spewing 100 times more lava than Earth does.

Io's tallest volcanoes are as high as Mount Everest. They are probably made of silicate rock resulting from volcanic eruptions similar to those on Earth. The largest volcanoes, such as Pele, named for the Hawaiian volcano goddess and apparently still volcanically active, eject volcanic material in huge umbrella-shaped plumes that rise to an altitude of 150 miles or more and

Figure 71 *The Jovian moon Europa showing a complex array of streaks that indicate the crust has been fractured and filled with materials from the interior.*

(Photo courtesy NASA)

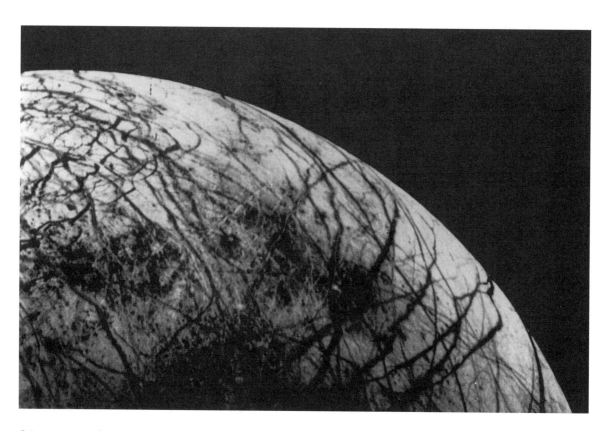

spread ejecta over areas as wide as 400 miles. The erupting hot lava is reminiscent of Earth some 4 billion years ago, when it was in a fiery turmoil. Therefore, Io might provide a glimpse into Earth's geologic youth, when its interior temperatures were much higher than they are today.

Saturn is similar in size and composition to Jupiter but has only one-third of its mass. Saturn possesses the strangest set of moons in the solar system (Fig. 73), with a total of 18. They range in size from an asteroid several hundred miles wide to larger than Mercury. All but the outer two moons have nearly circular orbits, lie in the equatorial plane (the same plane as Saturn's rings), and keep the same side facing their mother planet, as does Earth's moon.

The densities of Saturn's moons are less than twice that of water, indicating a composition of rock and ice. With an albedo of nearly 100 percent, Enceladus, the second major moon outward from Saturn, is the most reflective body in the solar system. Its icy surface is heavily dotted with meteorite craters, and long rilles appear to have been resurfaced by volcanic activity.

Hyperion appears to consist of fragments that recombined after collision with another large object had shattered it.

Rhea, the second largest of Saturn's moons, has a densely cratered surface (Fig. 74) similar to the highlands on Mercury and Earth's moon. Mimas has a heavily and uniformly cratered surface. Dione shares much of the same terrain features as Rhea and is also Saturn's second densest moon. Rhea and Dione are intensely cratered, while a single impact crater on Tethys punctures a hole two-fifths the diameter of the moon itself. Tethys has a branching canyon, 600 miles long, 60 miles wide, and several miles deep, spanning the distance between the north and south polar regions.

Titan is larger than Mercury and the only moon in the solar system with a substantial atmosphere, even denser than that on Earth. It is composed of compounds of nitrogen, carbon, and hydrogen and is believed to resemble

Figure 73 Saturn and its moons, from Voyager 1 in November 1980.

(Photo courtesy NASA)

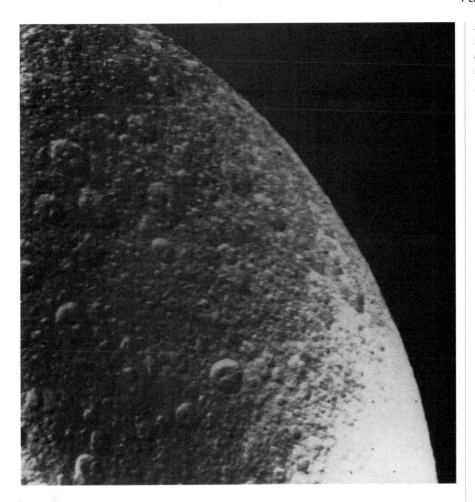

Earth's primordial atmosphere. Titan is therefore one of the best places to look for what Earth was like in its earliest stages of development. Titan is also the only known body besides Earth whose surface is partially covered by liquid, although its oceans are composed of liquid methane at temperatures of −175 degrees Celsius and its continents are made of ice.

Oberon and Titania, the largest and outermost of the Uranian moons, are both a little less than half the size of Earth's moon. Their surfaces, which are a fairly uniform gray, are rich in water ice. Bright rays, assumed to be clean buried ice brought to the surface, shoot out from around several meteorite craters. A few features on Oberon resemble faults, but the moon shows no evidence of geologic activity. Its surface is saturated with large craters, some more than 60 miles across. On the floors of several larger craters, volcanic activity

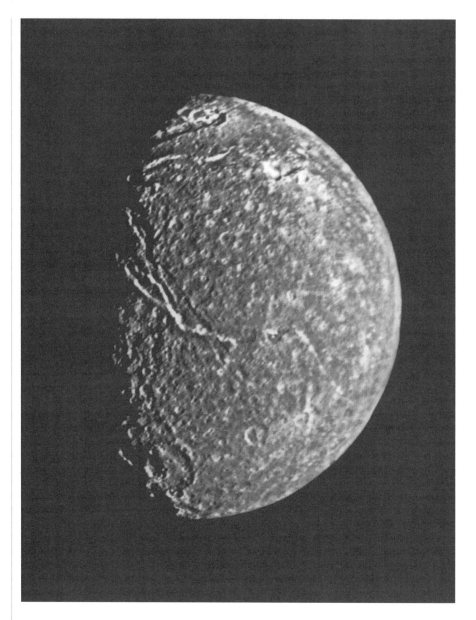

Figure 75 *Titania, the largest Uranian moon, shows a large trenchlike feature near the day-night boundary, which suggests tectonic activity.*

(Photo courtesy NASA)

has spewed forth a mixture of ice and carbonaceous rock from the moon's interior.

The surface of Titania (Fig. 75) bears strong evidence of tectonic activity, including a complex set of rift valleys bounded by extensional faults, where the crust is being pulled apart. Titania's surface was also heavily cratered in its early history. However, many larger craters were erased when the moon was resurfaced by volcanism that spilled huge quantities of water onto the crust. Some large craters might have disappeared because the soft, icy crust collapsed. When the water in the interior began to freeze and expand, the entire surface stretched. Meanwhile, the crust ripped open, enormous blocks of ice dropped down along the faults, and water upwelled through the cracks to form smooth plains.

Umbriel and Ariel are nearly equal in size and about one-third as large as Earth's moon. Umbriel is heavily cratered but lacks rays, which gives it a nearly bland appearance. This indicates the moon might be covered with a thick blanket of uniform, dark material composed of ice and rock. This feature also makes Umbriel the darkest of the Uranian moons, whereas Ariel is the brightest.

Ariel reveals evidence of solid-ice volcanism never before observed in the solar system. The ice rises through cracks in the crust and erupts on the surface, producing a most unusual landscape. Ariel is the least cratered of the major moons, indicating its surface was remade over and over again. Ariel was resurfaced by volcanic extrusion of a viscous mixture of water and rock that flowed glacierlike from deep cracks, similar to the flowage of lava from extensional faults on Earth.

A global network of faults crisscross Ariel and, in some places, they are tens of miles deep. A canyonlike feature called Brownie Chasma resembles a graben, a crack in the crust formed when the surface is pulled apart, causing large blocks to drop downward. The walls of the chasm are about 50 miles across, and the floor bulges up into a round-topped ridge about 1 mile high.

Miranda is perhaps the strangest world yet encountered. It might have acquired its bizarre terrain when it was shattered by comet impact and the pieces reassembled in a patchwork fashion. It is the smallest of the major moons of Uranus, with a diameter of only 300 miles. Yet for its modest size, it packs more terrain features than any moon (Fig. 76). The surface is covered with densely cratered rolling plains, parallel belts of ridges and groves 100 to 200 miles wide shaped like racetracks, and a series of chevron-shaped scarps. Huge fracture zones that encircle the moon creating fault valleys with steep, terraced cliffs that reach a height of 12 miles, cut the entire landscape.

Triton, the largest of Neptune's moons, appears to have been captured because it orbits tilted 21 degrees to Neptune's equator and rotates in a retrograde or opposite direction. This is the only large moon in the solar system known to do so. Triton is also the second most volcanically active body in the solar system behind Jupiter's moon Io. Since Triton's internal heat has long since vanished, the energy needed to power its volcanic eruptions remains a mystery.

Dark plumes seem to indicate highly active volcanism on a body that is thought to have been tectonically dead for 4 billion years. Apparently, gigantic nitrogen-driven geysers spew fountains of particles high into the thin, cold atmosphere and are strewn across the surface. Triton's geysers also might

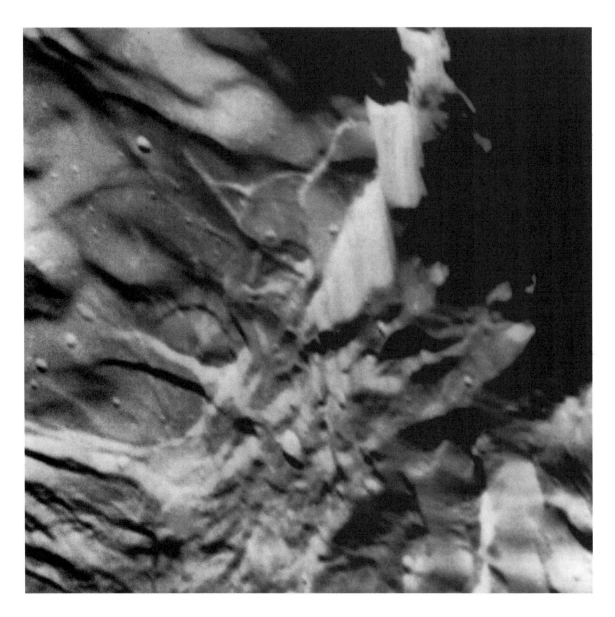

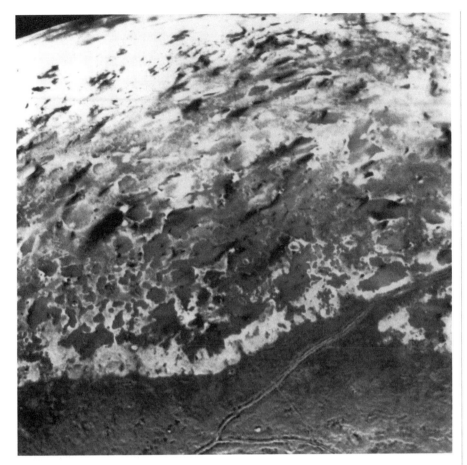

explain outbursts on comets, which cause an explosive brightening. Icy slush oozes out of huge fissures, and ice lava forms vast frozen lakes, providing one of the most bewildering landscapes in the solar system (Fig. 77).

After discussing meteorite impacts on the planets and their moons, the next chapter examines asteroids and meteorites.

5

ASTEROIDS
WANDERING ROCK FRAGMENTS

This chapter examines asteroids, the asteroid belt, meteors, and meteorites. The solar system is vast, with nine planets and their satellites stretching across billions of miles (Fig. 78). Between the orbits of Mars and Jupiter lies a wide belt of asteroids that contains 1 million or more stony and metallic irregular rock fragments. In addition, an estimated 2,000 asteroids larger than half a mile wide follow orbits that cross Earth's path around the Sun.

The asteroids formed either by individual accretion at an early age or by the breakup of larger bodies after collisions with other space debris. The combined mass of the asteroids equals about half that of Earth's moon. The original volume of asteroidal material was probably much greater, having been heavily depleted as asteroids rained down on the inner solar system. The burned-out hulks of comets passing close by the Sun might have replenished the asteroid belt.

THE MINOR PLANETS

Unlike planets and comets, known since ancient times, asteroids are a relatively recent discovery. According to the Titius-Bode Law, which defines the orbits of the planets around the Sun, a planetary body should have existed in the wide region between Mars and Jupiter. On January 1, 1801, while searching for the "missing planet," the Italian astronomer Giuseppe Piazzi discovered instead the asteroid Ceres, named for the guardian goddess of Sicily. It is the largest of the known asteroids, with a diameter of more than 600 miles. A subsequent search of the region uncovered several additional asteroids. Today, thousands have been tracked and cataloged.

Asteroids, from Greek meaning "starlike," were once thought to be the debris from a shattered Mars–sized planet. They are actually the remnants of a planet that failed to form due to a gravitational tug of war between Jupiter and Mars. Asteroids therefore offer significant evidence for the creation of planets and provide clues as to conditions in the early solar system. The majority of asteroids orbit the Sun within a confined area between the orbits of Mars and Jupiter called the main asteroid belt. Due to the proximity of Mars to the asteroid belt, its moons Phobos and Deimos (Fig. 79) might have been captured main-belt asteroids.

Most asteroids revolve around the Sun in elliptical orbits. Over the years, the paths of the major asteroids have been accurately plotted. Of the million or so asteroids with diameters of half a mile or more, some 18,000 have thus far been located and identified. The orbits of about 5,000 have been precisely determined. The orbits of some asteroids occasionally stretch out far enough to come within the paths of the inner planets, including Earth.

The three largest asteroids, Ceres, Pallas, and Vesta, are hundreds of miles wide and contain about half the total mass of the asteroid belt. Nearly 1,000 asteroids are larger than 20 miles across. Of these, more than 200 are larger than 60 miles wide. The volume of space in the main belt is so vast, however, that collisions among asteroids are infrequent and occur only over geologic timescales. When asteroids do collide, the impacts chip away at the surfaces, providing numerous fragments (Fig. 80) that often fall to Earth as meteorites. An asteroid might even masquerade as a comet with a tail of dust blasted from its rocky surface during collision with another asteroid.

Over eons, constant collisions between asteroids and the gravitational influence of Jupiter have depleted the main belt, clearing the asteroid zone of most of its original mass. As a result, those observed today are only remnants of the past. The existence of dust bands within the asteroid belt confirm that collisions among asteroids are an ongoing process. Often, the collisions erode the rocky surface material, exposing the more solid metallic cores. The various families of asteroids therefore result from the disruption of large parent bodies. The remaining fragments provide valuable information about the interiors of the asteroids.

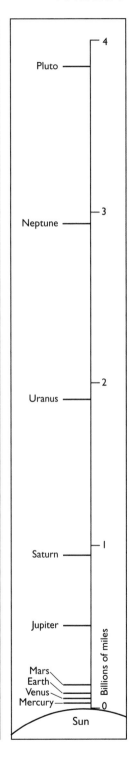

Figure 78   *The distances of the planets, in billions of miles.*

Figure 79 Mars's tiny moon Deimos might have been a captured main-belt asteroid.

(Photo courtesy NASA)

Figure 80 (1) A planetoid smaller than the Moon is (2) broken up by a giant collision, and (3) additional collisions yield asteroids that bombard Earth.

Not all asteroids reside within the main belt. An asteroidlike body has a far-ranging orbit that carries it from near Mars to beyond Uranus. One large object, called Chiron, lies between the orbits of Saturn and Uranus—an odd place for an asteroid.

An interesting group of asteroids, called Trojans, lies in the same orbit as Jupiter in two separate clumps, one preceding the planet and the other following it. The Trojan asteroids are probably as numerous as those in the main asteroid belt. Each of the two swarms of asteroids clusters about what

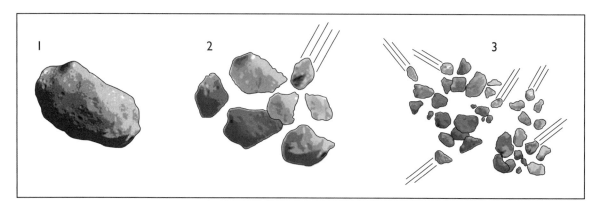

is called a Lagrange point. This is a region of gravitational stability where the centrifugal force of the asteroids' orbits counterbalances the gravitational pull of Jupiter and the Sun. However, some Trojans travel in orbits quite a distance from these stable regions and could possibly begin to wander into the inner solar system. Some 1,200 Trojans appear to have left their parent swarms and are in orbits that could take them within striking distance of Earth.

Asteroids are generally irregular in shape and rotate on their spin axis typically once every four to 20 hours. Due to collisions, which add rotational angular momentum randomly, small asteroids spin quite fast, whereas larger asteroids with higher moments of inertia spin more slowly. This trend curiously reverses for asteroids larger than about 75 miles wide, whose spin rates tend to rise with increasing size. Apparently, they have enough mass for gravity to keep them intact after collision, which spins the bodies more rapidly.

Most asteroids reflect varying amounts of sunlight as they rotate because of their irregular shapes. Binary asteroids, which orbit each other, have a pulsating brightness as they pass near Earth. They tend to flicker as one body passes in front of or behind its companion asteroid. These types of near-Earth asteroids are apparently quite common.

Many large, rapidly rotating and elongated asteroids appear to be gravitationally bound "rubble piles" that reassembled after being thoroughly shattered after a collision. The existence of rubble-pile asteroids also explains the close pairs of craters common on the planets. Perhaps as many as one-fifth of the large asteroids that cross Earth's path are orbiting pairs. They were possibly torn into two smaller bodies that orbit each other after passing close to Earth, creating a crater pair upon impact during the next encounter.

Asteroids are classified based on their composition. They fall into the three basic groups of primitive, metamorphic, and igneous. The distribution pattern suggests a steep temperature gradient and strong solar wind early in the formation of the solar system that altered the composition of the asteroids. Internal heating agents, such as radioactive elements along with impact friction, would have made a major contribution to the heating process.

Primitive asteroids, called P types, dominate the outer region of the main belt. They are primarily rich in carbon and water and represent unaltered remnant material from the formation of the solar system. Metamorphic asteroids, called C types, also contain abundant carbon and reside in the central region of the belt. They resemble primitive asteroids except for fewer volatile compounds and very little water. Increased heating, transforming some primitive asteroids into metamorphic ones, drove off these substances. The C-type asteroids, such as Pallas, are made up of more easily erodible materials than the stony or metallic asteroids. Therefore, their surfaces are ground down more finely than other types.

Igneous asteroids, called S types, are most frequently found in the inner region of the belt. They are apparently the source of the most common class of meteorites, known as the ordinary chondrites. These asteroids appear to have been strongly heated and formed from a melt with a complex mineralogy, similar in composition to Earth's mantle, consisting of olivine, pyroxene, and metal. A major source for the ordinary chondrites might be the 115-mile-wide asteroid Hebe, which lies in a zone of the asteroid belt where debris splashed off by impacts can fling meteorites toward Earth.

THE ASTEROID BELT

Long before Earth and the other planets formed around the Sun, a vast cloud of dust had begun to coalesce into asteroids. Many of these pieces of space debris were large and hot enough to erupt molten lava. However, the vast majority of asteroids were smaller chunks of cold rubble. Most meteorites showering down on Earth arrive from the main asteroid belt, a 250-million-mile-wide band of primordial debris inclined about 10 degrees with respect to the ecliptic, the plane of the solar system. The asteroids range in size from chunks of rock hundreds of miles wide down to less than one hundred feet wide. Small grains broken off asteroids become micrometeorites when striking Earth. Since the beginning, Earth, its moon, and the other planets and their moons have been peppered by meteorites originating from the asteroid belt.

The asteroid belt contains about 1 million pieces of solar system rubble larger than half a mile wide along with a substantial number of smaller objects. Zodiacal dust bands of fine material orbit the Sun near the inner edge of the main asteroid belt (Fig. 81). The ring is about 30 million miles wide and several hundred thousand miles thick, with Earth embedded in its inner edge. The debris appears to have originated from collisions between asteroids and from comet trails, comprising dust and gas blown outward by the solar wind.

Asteroids are leftovers from the creation of the solar system. Due to the strong gravitational attraction of Jupiter, they were unable to consolidate into a single Mars-sized planet in the region of the asteroid belt even though in theory one should be there. Instead, they originally formed several small planetoids up to 50 miles or more wide along with a broad band of debris in orbit around the Sun.

The largest asteroids might have been heated by radiogenic sources and melted from the inside out, thereby differentiating early in the formation of the solar system. Asteroids of the inner and middle belt underwent considerable heating and experienced as much melting as the inner planets. The molten metal in the asteroids along with siderophiles (iron lovers), such as iridium and osmium of the platinum group, sank to the interiors and solidified.

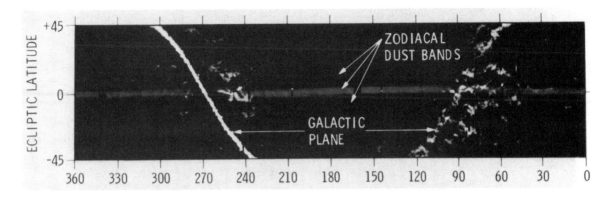

Figure 81 *The zodiacal dust bands are debris from comets and collisions between asteroids near the inner edge of the asteroid belt.*

(Photo courtesy NASA)

The metallic cores were exposed after eons of collisions among asteroids chipped away the more fragile surface rock. Then, breakup after collisions yielded several dense, solid fragments. Many asteroids contain a high concentration of iron and nickel, suggesting they were once parts of the metallic cores of planetoids that disintegrated after a collision with other planetary bodies.

The stony asteroids lying near the inner portion of the asteroid belt are less dense and have a high percentage of silica. The darker carbonaceous asteroids containing large amounts of carbon lie toward the outer region of the asteroid belt. Three classes of meteorites, the eucrites, howardites, and diogenites, are made of basalt originating from the surface rock of a single large asteroid.

Certain rare basaltic meteorites, the eucrites, might be pieces of the asteroid Vesta, the third largest in the solar system. Vesta is also one of the few objects in the main asteroid belt that contains a high concentration of pyroxene, a common igneous rock–forming silicate mineral. Vesta is one of the oldest asteroids, which assembled, melted, and partially cooled within 5 million years of the solar system's formation. However, a mystery remains as to how fragments from Vesta managed to make their way to Earth from the asteroid's location in the most distant reaches of the asteroid belt.

KIRKWOOD GAPS

A peculiar feature of the asteroid belt is the presence of large gaps, where few if any asteroids are found. Between the inner and outer regions of the asteroid belt are wide spaces called Kirkwood gaps, named for the American mathematician Daniel Kirkwood who discovered them in the 1860s. The six Kirkwood gaps are almost totally devoid of asteroids. If an asteroid falls into one of these spaces, its orbit stretches wide enough to make it swing in and out of the asteroid belt, bringing it close to the Sun and the orbits of the inner planets.

Each Kirkwood gap occupies a location that is a multiple of Jupiter's orbital period. For example, an asteroid orbiting in a nearly circular path in a Kirkwood gap about 2.5 astronomical units from the Sun (one astronomical unit, designated A.U., is the Earth's distance from the Sun or 93 million miles) will orbit three times for every one of Jupiter's orbits. After repeated passes at the same points in their orbits over a period of about 1 million years, the gravitational pull of Jupiter throws the asteroid into a highly elliptical orbit that can cross the path of Mars around the Sun. Consequently, an asteroid falling into a Kirkwood gap would not remain there very long, which explains the presence of the gaps in the first place.

The 2.5 A.U. Kirkwood gap might be responsible for a curious property exhibited by some meteorites. Apparently, twice as many stony meteorites, the most common type, fall to Earth during the afternoon than in the morning. This strange phenomenon has long been explained by assuming that more people are outdoors during the afternoon hours and therefore find more meteorites at that time. Yet farmers, who uncover the most meteorites in their fields, usually begin work early and thus are active throughout the day. Apparently, the meteorites follow orbits that cause them to intersect the afternoon hemisphere of Earth more than they do the morning hemisphere (Fig. 82).

The total mass of stony meteorites originating from the 2.5 A.U. gap that falls through Earth's atmosphere is about 100 tons annually. Most meteorites are lost because they disintegrate in the atmosphere or plunge into the ocean. Yet meteorites that do land on the ground appear to do so most likely in the afternoon.

Asteroids also occupying certain zones, known as resonances, within the outer part of the main belt are greatly influenced by Jupiter's gravity due to its close proximity. The giant planet's pull can dramatically elongate the orbits of these asteroids, causing their paths to cross the orbits of the inner planets. Another set of resonances in the main belt near Mars also cause some asteroids to escape into the inner solar system.

Figure 82   More stony meteorites fall on the afternoon hemisphere than on the morning hemisphere.

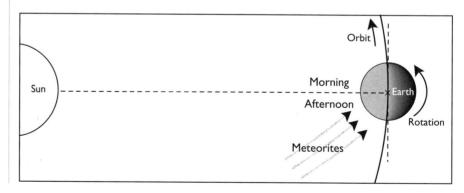

The Yarkovsky effect, named for the Russian engineer who discovered it a century ago, results from the way an asteroid absorbs and reradiates solar energy, acting as a sort of rocket engine. This effect provides enough energy to push a significant number of small asteroids less than 6 miles wide into resonances that can deliver them into the inner solar system. The phenomenon could explain why the smallest members of one family of asteroids, known as Astrids, have the widest range of orbits that could reach Earth.

METEROID STREAMS

Meteors, also called shooting stars or falling stars, are celestial objects that streak across the sky, exhibiting long tails of shimmering light. They are often seen at night and sometimes even during the daytime. Meteors flare up when they fall through Earth's atmosphere and are a common sight on clear nights away from bright city lights. Most meteors begin to flare up at an attitude of between 70 and 55 miles at the very reaches of the outer atmosphere. The meteoroids that produce meteor showers comprise tiny, fluffy dust particles derived from comets. High-altitude reconnaissance aircraft have actually collected some of these particles.

Meteoroids are rock or dust particles residing outside Earth's atmosphere in interplanetary space. They are small fragments sloughed off of comets or pieces of asteroids broken up by numerous collisions in space. This makes meteorite falls quite common due to the immense number of these fragments.

When a meteoroid falls through the atmosphere, usually between 70 and 40 miles altitude, it is heated by air friction. This makes the adjacent air molecules glow, producing a short-lived streak of light known as a meteor. The extreme heat generated by atmospheric friction ablates (peels off) the meteor's outer layers, whose incandescence produces a long, fiery tail generally lasting less than 1 second.

Only meteors of a certain size are sufficiently large to travel all the way through the atmosphere without completely burning up. A meteoroid landing on Earth's surface is therefore called a meteorite; the suffix "ite" designates it as a rock. Meteorites comprising rock or iron do not appear to originate from the meteoroid streams created by the tails of comets. They are instead fragments of asteroids chipped off by constant collisions.

Daily, thousands of meteoroids rain down on Earth, and occasional meteor showers can involve hundreds of thousands of tiny stones. Most meteors burn up on their journey through the atmosphere, and their ashes contribute to the load of atmospheric dust. Upward of 1 million tons of meteoritic material are produced annually, much of which remains suspended in the atmosphere, where it scatters sunlight and helps make the sky blue.

METEORITE FALLS

Meteorite falls are quite commonplace and have been observed throughout human history. Historians have often argued that a spectacular meteorite fall of 3,000 stones at l'Aigle in the French province of Normandy in 1803 sparked the early investigation of meteorites. However, this spectacle was actually eclipsed nine years earlier by a massive meteorite shower in Siena, Italy, on June 16, 1794. It was the most significant fall in recent times and spawned the modern science of meteoritics.

The ancient Chinese made the earliest reports of falling meteorites during the seventh century B.C. An interesting note is that Chinese meteorites are rare. To date, no large impact craters have been recognized in China. The first report of a meteorite impacting on the Moon was a flash witnessed by a Canterbury monk on June 25, 1178. Small asteroids account for every lunar crater less than 1 mile wide formed during the last 3 billion years (Fig. 83).

The oldest meteorite fall with material still preserved in a museum is a 120-pound stone that landed outside Ensisheim in Alsace, France, on November 16, 1492. The largest meteorite found in the United States is the 16-ton Willamette Meteorite discovered in 1902 near Portland, Oregon. It crashed to Earth sometime during the past million years and measured 10 feet long, 7 feet wide, and 4 feet high.

One of the largest meteorites actually seen to fall was an 880-pound stone that landed in a farmer's field near Paragould, Arkansas, on March 27, 1886. The largest known meteorite find, named Hoba West, weighed about 60 tons and was located on a farm near Grootfontein, Southwest Africa in 1920. The heaviest observable stone meteorite landed in a cornfield in Norton County, Kansas, on March 18, 1948. It dug a pit in the ground 3 feet wide and 10 feet deep.

When a meteor nears the end of its path through the atmosphere, it often explodes exhibiting a bright fireball called a bolide. One of the most impressive bolides resulted in the Great Fireball that flashed across the United States on March 24, 1933. Some bolides are bright enough to be seen during the daytime. Occasionally, their explosions can be heard on the ground, making a sound resembling the sonic boom of a jet aircraft. Thousands of bolides are estimated to occur around the world daily. However, because most explode over the ocean or in sparsely populated areas, they go completely unnoticed.

Of the more than 500 major meteorite falls that strike yearly, most plunge into the ocean and accumulate on the seafloor. For the great majority of meteorites that land on the surface, the braking action of the atmosphere slows their entry so they bury themselves only a short distance into the ground. Not all meteorites are hot when landing because the lower atmosphere tends to cool the rocks, which in some cases are covered by a thin layer

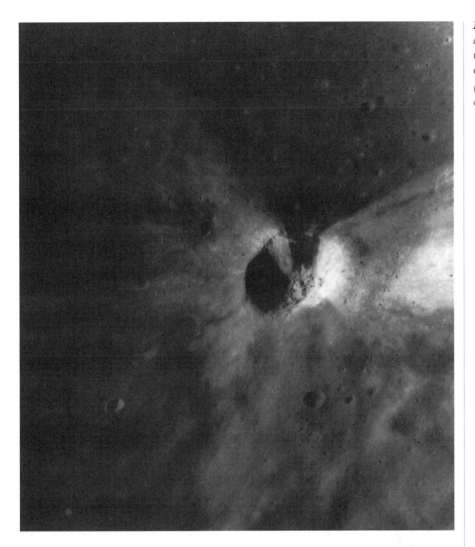

Figure 83 *A lunar crater more than 1 mile wide, showing an area of ejecta material.*

(Photo by H. J. Moore, courtesy USGS)

of frost. Meteorites can also cause a great deal of havoc as exhibited by the many examples of stones crashing into houses and automobiles.

The most easily recognizable meteorites are the iron variety, although they represent only about 5 percent of all meteorite falls. They are composed principally of iron and nickel along with sulfur, carbon, and traces of other elements. Their composition is thought to be similar to that of Earth's metallic core and might have once comprised the cores of large planetoids that disintegrated eons ago. Due to their dense structure, iron meteorites tend to survive impacts intact, and most are found by farmers plowing their fields.

Stony meteorites are the most common type, comprising some 90 percent of all falls. However, because they are similar to Earth materials and therefore erode easily, they are often difficult to find. The meteorites are composed of tiny spheres of silicate minerals in a fine-grained rocky matrix called chondrules, from the Greek *chondros* meaning "grain." Therefore, the meteorites that contain them are known as chondrites. Chondrules are believed to have formed from clumps of precursor particles when the solar system was emerging from a swirling disk of gas and dust within 5 million years of the Sun's birth.

The chemical composition of most chondrites is thought to be similar to the rocks in Earth's mantle. This has lead to speculation they were once parts of large planetoids that disintegrated early in the formation of the solar system. One of the most important and intriguing varieties of chondrites are the carbonaceous chondrites, which contain carbon compounds that might have been the precursors of life on Earth. The H types are primitive carbonaceous chondrites believed to have formed soon after the creation of the Sun and planets, and they comprise half of all chondrites.

One of the best hunting grounds for meteorites surprisingly happens to be on the glaciers of Antarctica (Fig. 84), where the dark stones stand out in stark contrast to the white snow and ice. When meteorites fall onto the continent, they embed themselves in the moving ice sheets. At places where the glaciers move upward against mountain ranges, the ice sublimates (evaporates

Figure 84   *The Antarctic peninsula ice plateau, showing mountains literally buried in ice.*

(Photo by P. D. Rowley, courtesy USGS)

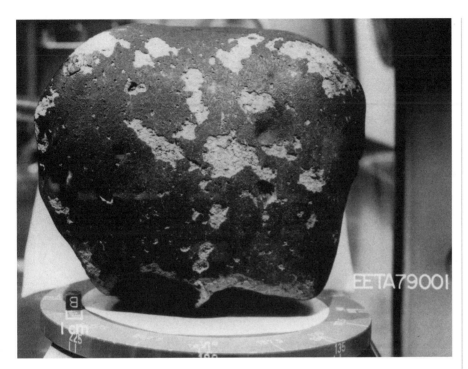

Figure 85 A meteorite recovered from Antarctica in 1981 and thought to be of possible Martian origin.

(Photo courtesy of NASA)

without melting), leaving meteorites exposed on the surface. Similarly, the glaciers on Greenland offer another indication of meteorites landing on the world's largest island.

Some of the meteorites landing on Antarctica are believed to have come from the Moon and even as far away as Mars (Fig. 85), where large impacts blasted-out chunks of material and hurled them toward Earth. A meteorite from the Allan Hills region of Antarctica comprised diogenite, a common type of basalt from the asteroid belt possibly impact blasted out of the crust of Mars. Organic compounds found on a Martian meteorite landing in Antarctica hint of previous life on Mars. A 40-pound meteorite that landed in Nigeria, Africa, in 1962 was identified as having been a piece of Mars ejected by a massive collision millions of years ago. It wandered in space for some 3 million years before finally being captured by Earth's gravity.

Perhaps the world's largest font of evidence of meteorites is the Nullarbor Plain, an area of limestone that stretches 400 miles along the south coast of Western and South Australia (Fig. 86). The pale, smooth desert plain provides a perfect backdrop for spotting meteorites, which are usually dark brown or black in color. Since the desert experiences very little erosion, the meteorites are well preserved and found just where they land. More than 1,000 fragments have so far been recovered from 150 meteorites that fell during the

Figure 86 *The*
Nullarbor Plain in
southern Australia is one
of the best hunting
grounds for meteorites.

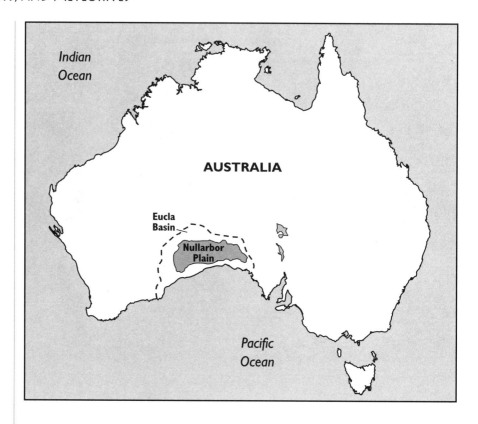

last 20,000 years. One extremely large iron stone, called the Mundrabilla Meteorite, weighed more than 11 tons.

EXPLORING ASTEROIDS

Meteorites are like space probes for free. They provide material samples, cosmic-ray traces, and geophysical data without the need for exploration by spacecraft. These data are important for studying the present as well as the early solar system. Thousands of meteorites have been cataloged, yet scientists cannot be sure where a single one actually came from.

The study of asteroids lying between Mars and Jupiter has long been cited as a major research goal. Of the 5,000 asteroids whose orbits are known (Table 6), only a few are deemed sufficiently interesting for planetary exploration. When the *Galileo* probe launched in October 1989 began its six-year odyssey to Jupiter, the spacecraft visited a pair of main-belt asteroids, Gaspra and Ida (Fig. 87), while in orbit within the inner solar system. The probe received a gravity assist from Venus and Earth before being hurled toward

TABLE 6   SUMMARY OF MAJOR ASTEROIDS

Asteroid	Diameter (miles)	Distance from Sun (million miles)	Type
Ceres	635	260	carbon-rich
Pallas	360	258	rocky
Vesta	344	220	rocky
Hygeia	275	292	carbon-rich
Interamnia	210	285	rocky
Davida	208	296	carbon-rich
Chiron	198	1,270	carbon-rich
Hektor	185X95	480	uncertain
Diomedes	118	472	carbon-rich

Jupiter. The complicated maneuver through the inner solar system (Fig. 88) was designed to provide observations of the terrestrial planets. On February 11, 1990, six months before the *Magellan* probe was due to arrive in orbit around Venus, *Galileo* began taking pictures of the cloud–shrouded planet.

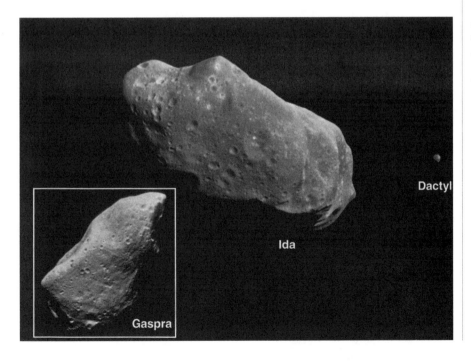

Figure 87 *A view of asteroid Ida with moon Dactyl and of asteroid Gaspra (insert), from Galileo.*

(Photo courtesy NASA)

Figure 88   *The flight path of* Galileo *to Jupiter, with flybys of Venus, Earth, and the asteroid belt.*

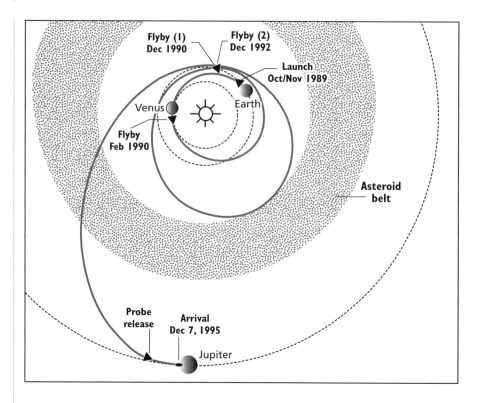

Figure 88   *The flight path of* Galileo *to Jupiter, with flybys of Venus, Earth, and the asteroid belt.*

On October 29, 1991, whizzing by at 18,000 miles per hour, *Galileo* rendezvoused within 900 miles of the 8-mile-wide asteroid Gaspra. The probe revealed that Gaspra was an S-type asteroid, resembling a stony meteorite in composition and possibly once part of a larger body. An unusual magnetometer reading suggested the tiny asteroid was metal rich and magnetized, making it the first small body in the solar system to possess a magnetic field. Gaspra is indeed primordial. The class of asteroids to which it belongs makes it a candidate source for the ordinary chondrite meteorites that so often fall to Earth.

After being hurled outward by Earth's gravity following its final flyby outbound toward Jupiter, *Galileo* whizzed past Ida at about 28,000 miles per hour on August 28, 1993. *Galileo* took a close look at and made many instrument readings of Ida, an S-type asteroid about 35 miles in length. More surprising was its tiny moon Dactyl, an orbiting rock less than 1 mile across that has been a companion of Ida for 100 million years or more.

On June 27, 1997, the *Near-Earth Asteroid Rendezvous (NEAR)* spacecraft came within 750 miles of the asteroid Mathilde. The asteroid was potato shaped, measuring some 40 miles long and 30 miles wide, with a mass of about 100 trillion tons and a density of only 1.3 times that of water. Analysis of images and data obtained during the flyby revealed that the carbon-rich, heavily cratered body was only half as dense as rocky asteroids. The asteroid appeared to be highly porous, suggesting it either was formed from loosely packed fragments or had been pulverized into a flying pile of rubble. The observable surface had five major craters ranging from 12 miles to 20 miles wide and as deep as 3.5 miles.

On February 14, 2000, *NEAR* orbited 15 miles above the asteroid Eros at a distance of some 196 million miles from Earth. Eros had escaped from the main asteroid belt shortly after it was formed in a catastrophic collision of two large asteroids. The asteroid's rapid spin suggests that the body once ventured close to Earth or Venus, which might explain its highly elongated shape. Data from *NEAR* revealed that meteorites had more heavily impacted the asteroid than was expected outside the main belt. The surface was almost completely covered with meteorite impacts, suggesting Eros has a very ancient surface.

One year later, on February 12, 2001, *NEAR* made a controlled crash onto Eros, successfully completing the first landing onto an asteroid. During its descent, the spacecraft's camera recorded close-up views of the asteroid that were sharper than any images previously taken. The probe landed just outside a saddle-shaped crater called Himeros. After landing, the transmitter continued to broadcast a strong signal. However, the camera located on the craft's landing side was disabled. Although blinded, onboard instruments could still send back precise data about the asteroid's composition.

The surface of large asteroids are believed to be covered with a loosely compacted layer composed of rocks shattered by meteorite collisions similar to the regolith on the Moon. Another possible mission would have as its primary target the asteroid Vesta, one of the largest and among the most interesting of the main-belt asteroids. A huge impact crater on the asteroid could be the source of many Earth-striking meteorites. The crater is 285 miles wide and 8 miles deep, roughly the size of Ohio, whereas Vesta itself is only 330 miles in diameter. Two possible additional flybys are also planned for Chaldaea and Helena, asteroids of differing compositions, with diameters of 75 and 45 miles, respectively.

By providing platforms for mounting telescopes and other scientific instruments, asteroids would make ideal space vehicles for exploring the solar system. Their use in this manner would greatly expand the knowledge of the solar system's planetary backyard. Asteroids might also someday be used as natural space stations from which to study the unexplored regions of the solar system.

Moreover, because of their weak gravitational attraction, asteroids would be much easier for spacecraft to land on, requiring only a docking maneuver. A telescope mounted onto an asteroid would offer astronomers a unique perspective without an interfering atmosphere. Asteroids would also make ideal space probes when fitted with instruments to gather information on much of the unexplained phenomena inside and outside the solar system.

MINING ASTEROIDS

Asteroids believed to have formed about the same time as the rest of the solar system are collection points for valuable minerals. Therefore, the mining of asteroids would be a potentially very economically profitable venture as space technology improves in the future.

An unusual geologic structure in Ontario, Canada, called the Sudbury Igneous Complex, is the world's largest single source of nickel and is also rich in copper and other valuable minerals. The geologic history of the complex has remained in question for more than a century. One of the most intriguing theories suggests the structure was formed when a large meteorite slammed into Earth 1.8 billion years ago. The affected area is roughly 20 miles by 30 miles and more than 1 mile thick. A similar ore body lies in the Bushveldt complex in South Africa, which also might have had a meteorite impact origin.

Historians believe that the first iron and, in some cases, perhaps the only iron used by ancient civilizations was derived from iron meteorites. Indeed, the word for iron is related in many languages to words like *star* or *sky*. In the ruins of the city of Ur in southwest Asia, pieces of a dagger dating to before 3000 B.C. was found containing more than 10 percent nickel that was meteoritic in origin. A variety of weapons have been made from iron meteorites since early times and were believed to give the bearer mystical powers.

Diamonds have also been found in minute crystals in stone meteorites. These grains are usually only of microscopic size and therefore are of no use industrially or ornamentally. One diamond from an iron meteorite found in Canyon Diablo, Arizona, near Meteor Crater measured roughly 0.03 inches by 0.01 inches. The diamonds are generally colorless, yellow, blue, or black. Diamonds might also be created by the high pressures and temperatures generated by the impact of a large meteorite.

Asteroids have been dubbed "flying gold mines." A single mile-wide body comprising high-grade iron and nickel along with other valuable ores could be worth as much as $4 trillion at today's prices. M-class asteroids, which consist mostly of iron and nickel, reside well outside the orbit of Mars. However, objects of this type also come near Earth's orbit and thus represent a vast potential for metal ore.

The asteroid Psyche appears to be very metal rich and might have once been the metallic core of a larger body that broke up in a collision. Therefore, Psyche is the largest known piece of semirefined metal in the solar system. Some asteroids that come within a short distance of Earth for capture might also contain rich deposits of gold and platinum.

Earth-crossing asteroids are of special interest. Space probes could possibly begin to visit these objects in the not-too-distant future. One suggestion is that astronauts or robot space vehicles might mine these asteroids for useful materi-

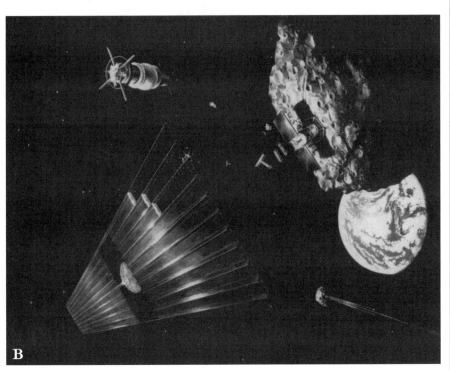

als. The asteroids would first be studied by space probes, which would use particle and laser beams to bombard the surface to determine its composition.

Several methods have been devised for using robots to process metals from asteroids in space. One idea developed by NASA is called the Asteroid Retrieval Mission (Fig. 89). It is designed to rendezvous with an Earth-approaching asteroid and bring it into the planet's orbit. The asteroid would then be mined for valuable ores, which would be hurled to Earth's surface using a system of mass drivers.

Mining asteroids in space could also provide the raw materials for building an array of space hardware, including huge spacecraft and space stations as well as facilities on other planets and their moons. These could be used as stepping stones across the solar system. Using them in this way might someday help the human race populate distant worlds.

After discussing asteroids, the next chapter examines other rogues of the solar system—namely comets.

6
COMETS
COSMIC ICE DEBRIS

This chapter examines Oort cloud and Kuiper belt comets and meteor showers. Asteroids and comets are distinctly different occupants of the solar system. Comets are hybrid planetary bodies, thought to consist of a stony inner core covered by an icy outer layer or a conglomeration of ice and rock. Comets are leftovers from the formation of the Sun and planets. They are often described as dirty snowballs composed of ice mixed with rock fragments. When passing near the Sun, the heat and solar wind deplete the comet's mass, as evidenced by its long, streaming tail. Ultimately, after repeated passes of the Sun, the comet burns out and masquerades as an asteroid.

Most comets follow orbits that are at oblique angles to the plane of the solar system (Fig. 90). Therefore, collisions with Earth are thought to be much less frequent than impacts by asteroids, which lie on the ecliptic. However, some 10 million comets are thought to cross the paths of the planets in the solar system. If a comet crosses Earth's orbit around the Sun and the planet enters its trail of dust, the night sky lights up with a meteor shower, providing one of the most impressive shows nature has to offer.

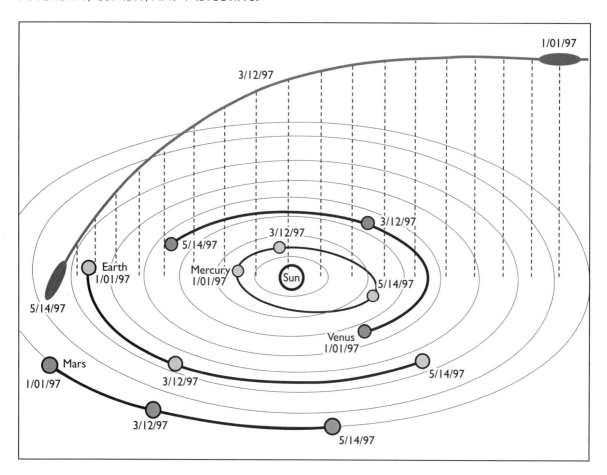

Figure 90 *Most comets follow orbits at oblique angles to the solar system.*

THE OORT CLOUD

Perhaps the first to speculate about the origin of comets was the fourth-century B.C. Greek philosopher Aristotle. He though that comets were clouds of luminous gas high in Earth's atmosphere. The first-century A.D. Roman philosopher Seneca disputed this notion and suggested instead that comets were heavenly bodies, traveling along their own paths through the firmament.

However, another 15 centuries passed before the Danish astronomer Tycho Brahe confirmed Seneca's hypothesis. He compared observations of a comet that appeared in the year 1577 from several locations around Europe and concluded that it was well beyond the Moon. In 1705, the English astronomer Edmond Halley compiled the first catalog of comets, which led him to make the discovery of the comet that bears his name.

Asteroids and comets represent the best-preserved samples of the early solar system. Comets are particularly important because of their primordial nature, which can provide clues about the processes that led to the origin of the Sun and all its orbiting bodies. Comets are also notoriously unpredictable, shedding gas and dust furiously at times, only to dry up and disappear altogether.

A shell of more than 1 trillion comets with a combined mass of 40 Earths surrounds the Sun about a light-year away. It is called the Oort cloud, named for the Dutch astronomer Jan H. Oort, who predicted its existence in 1950. Although the Oort cloud is heavily populated, comets are typically tens of millions of miles apart. The Oort cloud comets are so weakly bound to the Sun that randomly passing stars can readily change their orbits. About a dozen stars are thought to pass nearby the Oort cloud every million years. Occasionally, a star comes so close to the Sun it passes completely through the Oort cloud. These close encounters are sufficient to disrupt cometary orbits and send a steady stream of comets into the inner solar system lasting millions of years.

The currently active comets formed in the earliest days of the solar system and have since been stored in the celestial deep freeze of the Oort cloud. Temperatures this far away from the Sun are just a few degrees Celsius above absolute zero. Random gravitational jostling of stars passing nearby knock some outer comets in the cloud from their stable orbits and deflect their paths toward the Sun. One such body is Comet Halley (Fig. 91), which swings into view every 76 years. The comet should continue to exist for perhaps another 10,000 orbits or about half a million years before completely losing all its volatiles, at which time it will become a dried-up ball of space rubble.

Comets are characterized as flying icebergs mixed with small amounts of rock debris, dust, and organic matter. They are believed to be aggregates of tiny mineral fragments coated with organic compounds and ices enriched in the volatile elements hydrogen, carbon, nitrogen, oxygen, and sulfur. Comets might therefore more accurately be described as frozen mud balls with equal volumes of ice and rock.

Most comets travel around the Sun in highly elliptical orbits that carry them a thousand times farther out into space than the planets. Only when they swing close by the Sun, traveling at fantastic speeds, do the ices become active and outgas large amounts of matter. When the comets journey into the inner solar system, carbon monoxide ice vaporizes first and is replaced by jets of water vapor as the driving force behind the comets' growing brightness.

Comets flare up upon entering the inner solar system, where warmer temperatures ablate the outer covering of dirty ice and frozen gas. As they approach the inner solar system, water vapor and gases stream outward. This forms a tail pointing away from the Sun due to the outflowing solar wind. Sunlight reflecting off the dust and vapor trail reveals one or more tails possibly extending for 100 million miles or more. When approaching the Sun, the tails flow away from the comets. When leaving the Sun, the comets follow their tails (Fig. 92), which are often curved due to the motion of the comets relative to the solar wind.

As comets near the Sun's warming rays, some ice boils off into jets of water vapor that drag dust out along with them, causing the comets to flair up. Well before water ice warms enough to vaporize, it changes into a form that expels jets of volatile gases such as oxides of carbon trapped in the cometary ice. Because Oort cloud comets apparently preserved ices created near the primordial Sun, they tend to have a warm, low-density form of water ice, which might lead to noticeable differences in activity when the comets visit the inner solar system.

Because comets are so fragile, often depicted as a loosely held assembly of asteroid-sized snowballs, they tend to break up on their way toward the inner solar system. About 30 comets are known to have split their nucleus and due to the extra release of matter, this action tends to brighten them. Cracks and rifts on the icy nucleus set the stage for later fragmentation. The division of a cometary nucleus in this manner might give a fragment an orbit with a very low perihelion distance (closest approach to the Sun). A collision between two comets or between a comet and an asteroid might change the comet's orbit or break it up entirely.

Figure 91 Comet Halley, from the National Optical Astronomy Observatories.

(Photo courtesy NOAO)

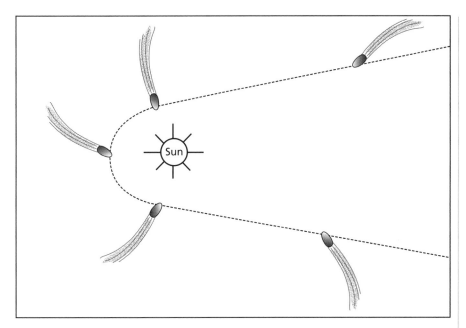

Hale-Bopp was one of the most impressive comets ever to visit the inner solar system. It made its closest approach to Earth 123 million miles away on March 22, 1997. Its icy core is 25 miles in diameter, or more than 10 times that of the average comet and 4 times larger than Comet Halley. It was the biggest and brightest comet in recent times and equal in size to the largest known comet, Schwassman-Wachmann 1, which follows a 15-year, nearly circular orbit just beyond Jupiter.

As it neared the Sun, Hale-Bopp began pumping prodigious quantities of matter into its 1-million-mile-wide coma, a gas envelope consisting of an extremely low density collection of gas and dust particles. A comet's brightness is generally determined by the amount of material that spews off as it enters the inner solar system. This material consists of dust, water vapor, and other substances that remain solid in the frigid reaches of the outer solar system but turn to gas as the comet is heated during close approaches to the Sun.

During Hale-Bopp's entry into the inner solar system, the comet released about 1 ton of carbon monoxide per second. This, in turn, liberated dust particles that spiraled off as the comet hurtled through space. Often, the comet brightened as it briefly expelled as much as 10 times more gas. Each expulsion probably represented the rotation of the comet as a particularly active part of its surface spun into view and then turned away.

Active areas erupted over the entire surface of the comet, presumably because it was close enough to the Sun for water to boil off its surface. The comet was recognized for its highly distinctive twin tails. The longer tail consisted of electrically charged atoms disbursed by the solar wind, whereas the shorter, curving tail contained dust grains. Hale-Bopp's last visit to the solar neighborhood was about 4,200 years ago when Babylonian astronomers probably first made its acquaintance.

THE KUIPER BELT

Following the discovery of the planet Pluto in 1930, astronomers began searching the wide heavens for a 10th planet circling the Sun. More than 60 years elapsed before the elusive planetary body was finally found well beyond Neptune. The mysterious object was only a few hundred feet across, much too small to be another planet. Since then, dozens of these cosmic bodies have been discovered, providing evidence for a band of comets inclined less than 5 degrees to the ecliptic. This ring of comets is called the Kuiper belt, named for the Dutch-American astronomer Gerard P. Kuiper, who predicted its existence in 1951. The comet belt contains a family of comets distinct from those of the Oort cloud.

The Kuiper belt lies much closer to the Sun than the Oort cloud but is still much farther than Pluto, which due to its odd orbit might itself be a captured comet nucleus, an asteroid, or a moon of Neptune gone astray after colliding with a comet. Pluto, whose tiny moon Charon appears to have been gouged out by a violent collision with a comet, might actually be the largest member of the Kuiper belt. Pieces of Pluto and its moon called Plutinos blasted off by ancient impacts not only roam the comet belt, but some of them might visit the inner solar system and fall to Earth.

One former member of the Kuiper belt too far away to be seen from Earth is believed to be a huge comet about 250 miles wide. The comet apparently has a strange orbit that takes it far outside the Kuiper belt and to within several hundred million miles of Neptune. Such an oblong orbit signifies the object came under the influence of a massive body and escaped the Kuiper belt. Alternatively, an unseen planet about the size of Mars could have influenced the comet's orbit. Such a body could have coalesced in the outer solar system from the same debris that formed Neptune, Uranus, and the cores of Saturn and Jupiter.

Certain comets seem to plunge inward toward the Sun from the outer reaches of the solar system on a regular schedule. Many of these comparatively small objects composed of ice and rock periodically provide spectacular appearances as they near the Sun, which drives off dust and gas to form luminous halos and long, streaming tails (Fig. 93). These active comets are apparently new members of the inner solar system. For if they were created during the formation of the rest of the planets, 4.6 billion years of solar heating would have completely driven off their volatile substances, turning them into inactive rocky nuclei.

Astronomers divide comets into two groups according to the time they take to orbit the Sun, which is directly related to their distance from the Sun. Long-period comets originating from the Oort cloud with orbits of up to 200 years or more enter the interplanetary region from random directions, as would be expected for bodies arriving from a spherical repository. Conversely, short-period comets, which remain active for tens of thousands of years, normally occupy smaller orbits of less than 200 years and are tilted only slightly to the ecliptic. They therefore fall toward the Sun from nearly the same plane as the orbits of the planets.

Short-period comets are further subdivided into two groups. The Jupiter-family comets, with periods of less than 20 years, have orbits whose planes are typically inclined no more than 40 degrees to the ecliptic. The intermediate-period or Halley-type comets, with periods between 20 and 200 years, enter the planetary region randomly from all directions. The intermedi-

Figure 93 *Comet Ikeya-Seki, from the U.S. Naval Observatory.*

(Photo courtesy NASA)

ate-period and long-period comets appear to originate from the Oort cloud, whereas the Jupiter-family comets appear to come from the Kuiper belt.

The comets might have originally traveled in large, randomly oriented orbits similar to the long-period comets. They might have been diverted into a flat ring by the gravitational attraction of the large outer gaseous planets, principally Jupiter. However, Jupiter's gravity appears too weak to transform long-period comets from the Oort cloud efficiently into short-period ones. Therefore, the probability of gravitational capture seems too remote to account for the large number of short-period comets that tend to lie in planes close to the ecliptic.

Apparently, the disk of the solar system does not abruptly end at the orbits of Neptune and Pluto, which swings in and out of Neptune's orbit but avoids collision by a quirk of orbital geometry. This feature lends credence to the idea that Pluto, whose orbit tilts 17 degrees to the ecliptic (Fig. 94), might

Figure 94 *The orbits of the planets.*

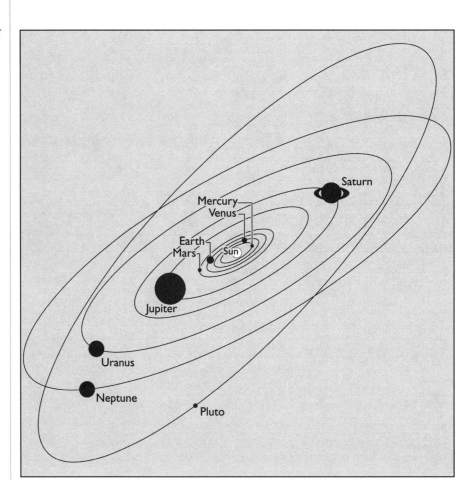

have been a captured asteroid, comet, or moon of Neptune knocked out of orbit by a passing comet. The two planets vie with each other as the most distant from the Sun.

A belt of residual material left over from the creation of the planets formed beyond Neptune and Pluto. The density of this matter in the outer region was insufficient for large planets to have accreted. Instead, objects the size of asteroids developed. These scattered remnants of primordial material were so far from the heating rays of the Sun that their surface temperatures remained quite low. Therefore, their composition is thought to be mostly water ice and various frozen gases similar to the nuclei of comets.

The known members of the Kuiper belt share common features. They lie beyond the orbit of Neptune, travel in orbits only slightly inclined to the ecliptic, and range in diameter from about 50 to 250 miles. The total population within the Kuiper belt is estimated to be at least 35,000 objects larger than 50 miles wide, giving a total mass hundreds of times greater than the asteroid belt. The Kuiper belt apparently maintains enough mass to provide all the short-period comets that have ever existed.

The Kuiper belt serves as a supply source for the rapidly depleted short-period comets. Neptune's gravity slowly erodes the inner edge of the belt, launching objects from within that zone into the inner solar system, where they ultimately burn up or collide with the Sun, the planets, or their moons. Perhaps the most spectacular collision was when fragments of Comet Shoemaker-Levy 9 plunged into Jupiter in July 1994 (Fig. 95), sending up huge fireballs. Others might get caught in a gravitational slingshot that ejects them out of the solar system and into interstellar space, never to be seen again.

SUN-GRAZING COMETS

One of the great riddles of the solar system is why so many comets collide with the Sun. Some comets unable to escape the Sun's powerful gravitational attraction as they enter the inner solar system plunge straight into it. Their small size appears to have little effect, however, except producing perhaps a momentary slight brightening of the Sun's corona, the outermost portion of the solar atmosphere (Fig. 96).

Between 1979 and 1984, the U.S. Air Force satellite *Solwind* discovered five new comets, named Solwind 1 to 5, all of which passed close by the Sun but failed to reappear on the other side. Either the comets collided with the Sun or were destroyed by its intense heat as they passed close by. They were members of a family of Sun-grazing comets, which have defied every attempt at explanation.

The orbits of Sun-grazing comets are retrograde (in opposite direction of other bodies in the solar system), inclined about 140 degrees to the plane of the solar system, and appear to come from about 17 billion miles away. Then unexplainably, their orbits are perturbed to such an extent the perihelion distances become less than the solar radius. When the comets enter the inner solar system, they cannot escape the strong gravitational pull of the Sun and are pulled into it.

Because Sun-grazing comets do not originate from the Oort cloud, they are not perturbed by the gravity of a passing star as are most other comets. Also, because their orbits are steeply inclined to the ecliptic, the comets are not influenced by the gravitational attraction of the large planets. Perhaps, a hole opening up on one side of the comet allows gases to spew out, similar to a rocket engine, which changes its orbit sufficiently to steer it into the Sun.

In 1991, Comet Machholtz came within about 6 million miles of the Sun's surface, close enough to broil it. The 5-mile-wide comet is in an

Figure 95 Comet Shoemaker-Levy 9 collided with Jupiter in July 1994, producing massive plumes in the planet's atmosphere.

(Photo courtesy NASA)

Jupiter, May 18, 1994 (Before)

Jupiter, July 22, 1994 (After)

Comet P/Shoemaker-Levy 9

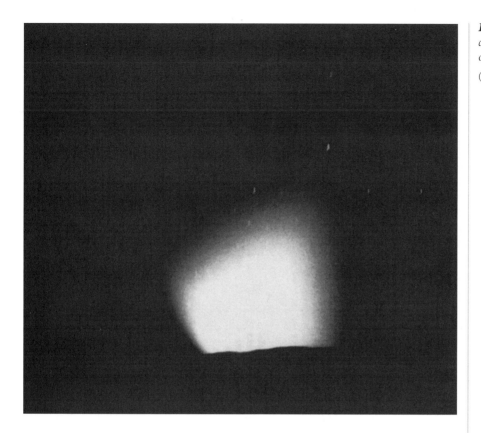

Figure 96   *The solar corona from* Apollo 15 *on July 31, 1971.*

(Photo courtesy NASA)

unusual orbit that gradually takes it closer to the Sun every 5.3 years. Its short orbit is causing the comet to spiral slowly into the Sun. Perhaps in a few decades, it will be unable to escape the Sun's powerful grasp and burn up in a cosmic flash.

As comets approach the Sun, they brighten and become redder. During the Middle Ages, people believed that comets with the color of blood were bad omens. Some comets might also show tinges of yellow or yellowish red, particularly if they were unusually bright. The brightness is determined by the albedo (Fig. 97) or reflectance value of the Sun's rays, which relies on the color and texture. Dirtier, redder surfaces produce a low albedo, whereas bluer, icier surfaces yield a high albedo.

Furthermore, the eruption of jets of fresh material from the nucleus can lead to sporadic color changes, which occur during the injection of icier material, turning the coma bluer. As the comet nears the Sun, it heats up and becomes dustier and consequently redder. The brilliant daylight Comet 1910 I was the last blood–red comet seen with the naked eye in the 20th century.

Much superstition surrounds comets. Since the beginning of recorded history, people have imbued comets with otherworldly significance. More

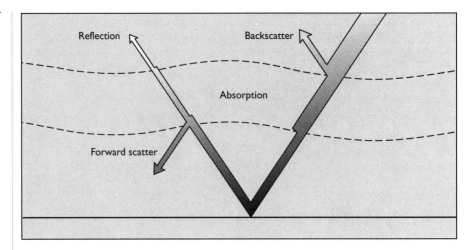

Figure 97 *The effect of the albedo on incoming solar radiation.*

than 2,500 years ago, the Babylonians thought comets were omens of floods and famines. The Carthaginian general Hannibal committed suicide in 182 B.C. supposedly after learning a newly spotted comet had foretold his death. A thousand years later, prophets used the visit of a comet to determine when Charlemagne would be crowned king of Germany. They viewed a later comet as a harbinger of his demise.

COMETS TURNED ASTEROIDS

Many Earth-crossing asteroids, called Atens, Apollos, and Amors, might have originated as comets. Throughout the eons, their coating of ices and gases has been eroded away by the Sun, exposing hunks of rock. These bodies are not confined to the main asteroid belt as are the great majority of known asteroids. Instead, they approach or even pass within the orbit of Earth (Fig. 98). Usually, inside a few tens of millions of years, the Apollos either collide with one of the inner planets, Earth included, or are flung out in wider orbits after a near miss.

Dozens of Apollo asteroids have been identified out of a possible total of perhaps 1,000. Most are quite small and were discovered only when approaching Earth. Many of these Earth-crossing asteroids do not originate from the asteroid belt but are believed to be comets that have exhausted their volatile materials after repeated close encounters with the Sun and are unable to produce a coma or tail. Inevitable collisions with Earth and the other inner planets are steadily thinning out their ranks, requiring an ongoing source of new Apollo-type asteroids either from the asteroid belt or from the contribution of burned-out comets.

For a comet to evolve into an asteroid, it must migrate from the Oort cloud and somehow enter a stable orbit in the inner part of the solar system. Meanwhile, cometary activity diminishes to the point that the comet becomes a burned-out hulk composed mostly of rock. Should a comet encounter one of the large planets such as Jupiter and become trapped in a short-period orbit, its path around the Sun would rarely be stable. Soon it would reencounter Jupiter and be flung back out into deep space, possibly escaping the solar system entirely.

Once a comet establishes a stable short-period orbit, it passes repeatedly near the Sun. Each time it encounters the Sun, it loses a few feet of its outer layers. The solar wind permanently removes the gas and dust, but the heavier silicate particles are pulled back into the nucleus by the comet's weak gravity. Gradually, an insulating crust forms to protect the icy inner regions of the nucleus from the Sun's heat to such an extent that the comet ceases its outgassing (Fig. 99). Thus, the comet can masquerade as an asteroid and even possess many of an asteroid's surface features.

The asteroid Chiron, which might provide a link between comets and asteroids, has assumed an orbit between Saturn and Uranus, a strange place, indeed, for an asteroid. Although first thought to be an unusual asteroid, Chiron is now firmly established as an active comet with a weak but persistent coma. It is about 112 miles in diameter or almost 20 times larger than Comet Halley. Its eccentric orbit carries the object inside the orbit of Saturn, whose interaction greatly destabilizes Chiron's path around the Sun. In a few million years, the gravitational influences of the giant planets could radically alter Chiron's orbit, either ejecting it out of the solar system or moving it toward

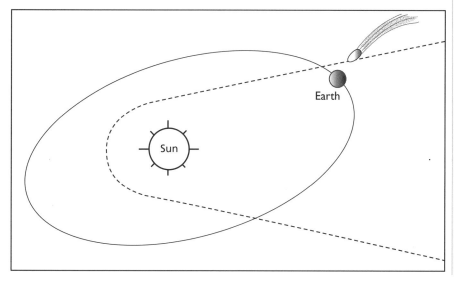

Figure 98 Many comets have orbits that cross Earth's path.

Figure 99 The life cycle of a comet. When it is young, fresh ice dominates the surface. At middle age, the comet develops an insulated crust. During old age, the crust becomes thick enough to cut off all cometary activity.

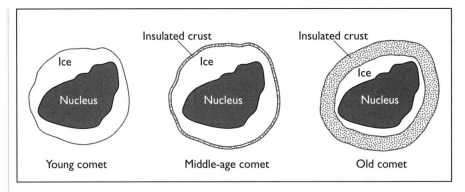

Jupiter. In the latter case, it might be injected into a short-period orbit that would bring it close by the Sun.

As it made its closest approach to the Sun in November 1987, Chiron might have revealed its true identity. It seemed to brighten much more rapidly than what would be expected of a bare, rocky object. A hint of a coma around Chiron along with its cometary activity has led to the suspicion it is a huge comet. Yet, Chiron appears too wide to be a comet nucleus. However, since Chiron does possess a coma, it could indeed be a very large comet.

METEOR SHOWERS

The nucleus of a comet consists of a core of rocky material surrounded by water ice and various frozen gases, interspersed with dust particles. Thus, comets are often depicted as dirty snowballs composed of ice and dust. As the frozen material at the surface of a comet nucleus evaporates when heated by the Sun, it turns into a gas that envelops the nucleus in an atmosphere called a coma, which extends tens of thousands to a million or more miles across.

The outward-streaming gases blow the dust previously trapped in the ice away from the comet's head in the direction opposite the Sun, thereby forming a long dust tail. All matter in the tail as it streams backward from the comet's head gradually fills the space around the comet's orbit, which is generally highly elliptical or cigar shaped. As Earth travels in a nearly circular orbit around the Sun, it passes through the dusty path of a comet (Fig. 100). As a result, the dust particles rain down through the atmosphere, producing a shower of meteors.

A meteor shower is a large number of meteors arriving from the same direction seen during a short interval ranging from a few hours to a few days. The huge collection of meteoroids orbiting the Sun is called a meteoroid stream, which produces a meteor shower when Earth passes through it. Apparently, a clear connection exists between the orbits of certain comets and the orbits of various meteoroid streams. For example, two annual meteor showers are associated with Earth passing through both legs of the dusty orbit of Comet Halley. They are the Eta Aquarid meteors, appearing in early May, and the Orionids, appearing around October 21. The suffix "id" when attached to the end of a name means the "daughters of."

The most prominent showers are the Perseids, appearing yearly around August 12, and the Geminids, appearing four months later around December 14. The greatest meteor shower of the 20th century was the Draconid on October 9, 1940. The most prominent meteor shower of the 19th century was the Leonid on November 13, 1833.

The parent comets for some meteor showers have yet to be identified. The parent comet for the Geminid meteors is thought to be an object named Phaethon. Its orbit, which takes it once around the Sun every 17 months, is almost identical with that of the Geminid meteoroid stream. However, Phaethon lacks any hint of an atmosphere or tail. Since other meteoroid streams are definitely associated with comets and none are associated with asteroids, Phaethon would appear to be a dead comet.

Asteroid collisions might also produce a sufficient number of small particles of debris to resemble a meteoroid stream. At aphelion, Phaethon lies deep inside the asteroid belt, where a collision with another asteroid could chip off enough material to produce the Geminids. However, such a collision would only be a one-time event. Therefore, the stream would not last as long as one created by the debris shed by a comet. A comet stream can persist for upward of millions of years as a comet discards its outer coating each time it passes by the Sun.

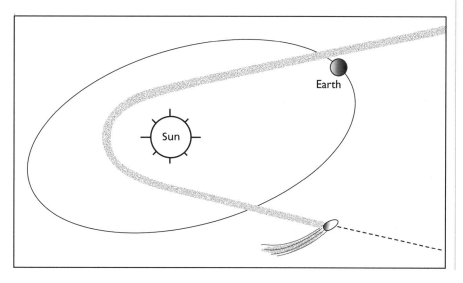

Figure 100 Meteor showers occur when Earth intercepts the dusty path of a comet.

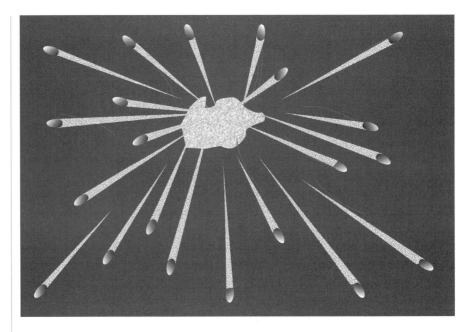

This material shed from comets becomes meteors when it intersects Earth's atmosphere. Although meteors streaking through the atmosphere are actually traveling in parallel lines, an observer on the ground perceives them as diverging from a central point in the sky, called the radiant (Fig. 101). The meteor shower is named after the constellation in which the radiant lies. For example, the Geminids appear to arrive from the constellation Gemini.

Due to the angle at which the meteors strike Earth's atmosphere, some meteor showers are visible only just before dawn. The dawn side of Earth faces the forward direction of Earth's orbit, enabling it to sweep up more meteors. Furthermore, many meteor showers can be seen only at certain latitudes. For example, the Eta Aquarid meteor shower is best observed in the Southern Hemisphere at a latitude of about 14 degrees south.

Radar systems tracking meteors have revealed that these small celestial bodies that rain down on Earth appear to be pebblelike objects plunging through the atmosphere. Yet satellite images of the sunlit side of the atmosphere imply that more massive clumps of material might be involved. These might be dirty, snowball type pieces of a comet, which shed their outer layers in the upper atmosphere, possibly releasing small pebbles trapped inside.

The *Dynamics Explorer* satellite launched in August 1981 detected holes apparently punctured through the atmosphere by meteors. The satellite was originally designed to observe Earth's dayglow, which is sunlight absorbed and reradiated by oxygen atoms in the atmosphere from a layer between 100 and 200 miles altitude. Dark spots on the satellite's ultraviolet imagery were believed to result from comets entering the atmosphere (Fig. 102). Each hole spread out in a fashion similar to the expansion of a drop of dye in a bowl of water. The effect caused the dayglow intensity to drop sharply over an area of about 1,000 square miles. Then over the next few minutes, the dayglow intensity returned to its normal value.

The number of dark spots doubled when a meteor shower passed through the atmosphere, which seemed to confirm the existence of the holes. Apparently, the meteors opened up the holes by a chemical reaction that reduced the number of free oxygen atoms responsible for the dayglow in the ultraviolet wavelength. The mass of the meteors would therefore have to be much larger than most. Perhaps they were minicomets made up almost entirely of water ice that is mostly empty space like snow. Meteoroids could also account for some of the dark spots in the ultraviolet images.

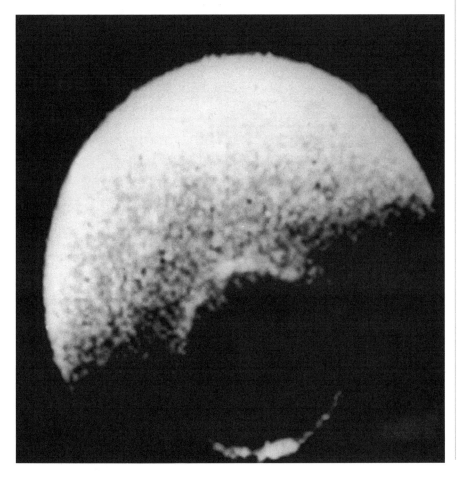

Figure 102 *An ultraviolet image of Earth's atmosphere from the* Dynamic Explorer *satellite. Dark spots are believed to be the result of small comets entering the atmosphere. The oval structure at the bottom is the aurora borealis.*

(Photo courtesy NASA)

Each day, 25,000 small, fluffy comets appear to bombard Earth, as indicated by signs of cometary water vapor in an extremely dry portion of the atmosphere at about 50 miles altitude. Small comets entering the upper atmosphere that vaporize due to air friction in less than 20 minutes apparently cause the bursts of water vapor. As many as 20 of these house-sized fluffy snowballs are thought to hit the atmosphere every minute, thereby continuously adding water to the planet. Because the comets must approach Earth from behind, they appear most frequently between midnight and noon.

The minicomets are believed to be about 40 feet wide and contain some 100 tons of water. They apparently swarm midway between Earth and the Moon. To avoid lighting up on entering the atmosphere, the minicomets probably orbit the Sun in the same direction and near the same plane as Earth. These strange objects might have escaped discovery out in space because they are cloaked in a thin layer of black organic material that insulates the ice against the Sun's heat, which prevents the formation of a coma and tail and hinders telescopic detection.

By their shear numbers, the minicomets would be contributing substantially to the amount of water and other meteoritic material added to the atmosphere. Furthermore, if the minicomets have been arriving at the same rate, approximately 1 million per year, over the age of Earth, they would provide enough water to fill the entire ocean.

COMETARY EXPLORATION

The first spacecraft to investigate the inner workings of a comet was the *International Cometary Explorer* (*ICE*), which plunged through Comet Giacobini-Zinner's tail on September 11, 1985, seven years after launching. The spacecraft, originally designed to study the solar wind, was maneuvered to approach within about 5,000 miles of the comet's nucleus. A pass this close to the nucleus was necessary to obtain critical information on the comet's interactions with the Sun's outpouring of protons, electrons, and magnetic field lines in the solar wind racing through the comet at upward of 300 miles per second.

When the spacecraft was some 80,000 miles from the comet's tail, it crossed a boundary zone called the bow shock. This is the shock wave generated when the ionized body of the comet encounters the solar wind, causing plasma particles to veer away. Upon entering the comet's tail, the probe found abundant water ions but fewer-than-expected carbon monoxide ions in the coma. Magnetic field lines captured from the solar wind were draped behind the inner coma, which was some 20,000 miles across, while isolated bundles of field lines pulled free of the tail as though being shed by the comet.

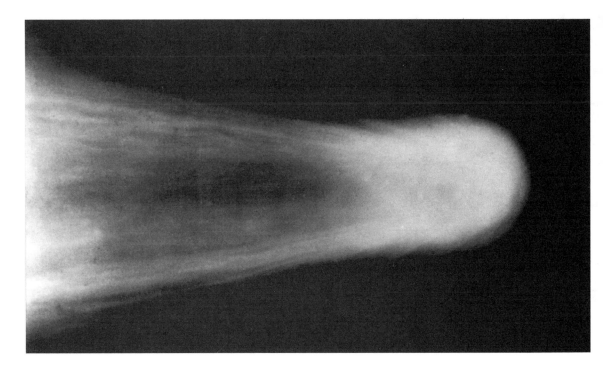

The spacecraft also detected dust particles, which vaporized and became ionized. Even when deep inside the coma, the spacecraft was hit by cometary grains only once every few seconds. The space probe, traveling at a speed of about 50,000 miles per hour, took about 20 minutes to cross the comet's plasma tail, which was well over 300,000 miles long. Scientists originally feared that a coating of comet dust would completely disable the probe. However, the little spacecraft came out of the comet's tail completely unscathed and went on to become one of six others to rendezvous with Comet Halley (Fig. 103).

The European Space Agency's *Giotto* spacecraft flew within 375 miles of the nucleus of Comet Halley in March 1986. The comet's nucleus appeared to resemble a 10-mile-long peanut encrusted with a layer of dust upward of tens of feet thick. The nucleus was found to contain simple organic molecules composed of hydrogen, carbon, nitrogen, and oxygen, which formed dark, tarlike substances. Between 80 and 90 percent of the mass of the comet consisted of dust and frozen water, a composition typical of the solar system's starting materials.

Nearly all the gas and dust forming the comet's tail erupted in powerful jets from only a few points on the nucleus. Rock-forming elements were found in Halley's dust in the same proportions as in chondritic meteorites— the most primitive of all meteorites. From the spiraling of dust jetting out of the nucleus, Halley appeared to be slowly rotating about one revolution a

Figure 103 *Comet Halley on its closest approach to Earth in 1986.*

(Photo courtesy NASA)

139

week. A similar side jet produced a distinctive blue tail on Comet Hale–Bopp when it approached the Sun in March 1997. NASA launched a series of small rockets to study the nucleus and tail in an attempt to reveal some of the comet's hidden secrets

The *Deep Space 1* spacecraft was originally built as a test bed for advanced sensors to be used on future NASA missions. However, it was redirected to fly to about 1,300 miles from a comet in September 2001. It took infrared images of the mysterious body far out in space to study the origins of the solar system.

After discussing comets and related phenomena, the next chapter takes a look at meteorite craters and impact structures scattered throughout Earth.

7

METEORITE CRATERS
FORMATION OF IMPACT STRUCTURES

This chapter examines meteorite impact cratering and related phenomena. Scattered throughout the world are wide circular features that appear to be impact structures created by the sudden shock of a large meteorite striking Earth's surface. The structures are generally circular or slightly oval in shape and range in size up to 100 miles or more in diameter. Some meteorite impacts are quite apparent, whereas most are only veiled outlines of impact structures. They might present themselves as an annular disturbed area, with rocks altered by powerful shock effects, requiring the instantaneous application of high temperatures and pressures like those deep in Earth's interior.

The search for ancient meteorite craters is severely hampered by vigorous erosional agents, which have long since erased all but the faintest signs. Meteorite craters on the Moon and the inner planets as well as the moons of the outer planets are quite evident and numerous. Many more meteorites likely hit Earth than the Moon because of its larger size and stronger gravitational attraction. Yet the Moon retains a better record of terrestrial impact cratering (Fig. 104). Fortunately, several remnants of ancient meteorite craters can

Figure 104   *Earth's*
moon, showing numerous
impact craters.

(Photo by H. A. Pohn,
courtesy USGS)

still be found on Earth, suggesting it was just as heavily bombarded as the rest
of the solar system. The evidence hints that impact cratering is an ongoing
process, and the planet can expect another major meteorite impact at anytime.

CRATERING RATES

Planetary science strives to compare the geologic histories of the planets and
their moons by establishing a relative time scale based on the record of impact
cratering. The rate of lunar impact cratering was similar to that on Earth
because both were subjected to the same intensity of asteroidal and cometary
impacts throughout the 4.6 billion years of geologic time. The cratering rate
of the Earth-Moon system was perhaps 100 times higher in the earlier years
than the cratering rate in more recent geology. Because the Moon lacks active

weathering agents that destroy most craters on its neighboring planet, it preserves a much better record of impact cratering. Although Earth's greater gravity captured more asteroids and comets than its satellite, the average crater on the Moon is preserved hundreds of times longer.

The form of the lunar craters changes as their size increases. Small craters exhibit a clean, bowl-shaped interior and a sharp rim crest. Larger craters more than 10 miles in diameter contain central peaks and are surrounded by a ring of terraces. These mark the head scarps of great blocks of rock that slid into the crater cavity shortly after formation. In giant craters, more than 100 miles wide, the central peak is replaced by an interior ring of mountains, and the impact structure is encircled by one or more inward-facing scarps.

Generally, the older and more stable the surface geology, the greater the number of craters. The heavily cratered lunar highland is the most ancient region of the Moon and contains a record of intense bombardment from around 4 billion years ago. Afterward, the number of impacts rapidly declined. The impact rate has remained relatively low and steady due to the depletion of the asteroid belt and other sources of large meteorites. Had the impacts continued on Earth at a high rate since the beginning, the evolution of life would have been substantially altered, if not halted altogether.

The cratering rate apparently differs from one part of the solar system to another. Asteroid and comet cratering rates along with the total number of craters suggest the average rates over the past few billion years were similar for Earth, its moon, and the rest of the inner planets. However, the cratering rates for the moons of the outer planets might have been substantially lower due in part to their greater distances from the main asteroid belt. Nonetheless, the size of the craters in the outer solar system (Fig. 105) is comparable to those of the inner planets.

The rate of cratering for Earth's moon and Mars was comparable, except on Mars erosional agents such as wind and ice erased most of its craters (Fig. 106). The southern hemisphere on Mars is rough, heavily cratered, and traversed by huge channellike depressions where massive floodwaters might have once flowed. Indeed, what appears to be ancient stream channels flowing away from meteorite craters seems to indicate that Mars once had liquid water. Most of the water is now locked up in the Martian polar ice caps or frozen underground.

On the Moon, the dominant mechanism for destroying craters is other impacts. Due to a high degree of crater overlap, placing the craters in their proper geologic sequence is often difficult. The impact rate for Mars might have actually been higher than the Moon, possibly because the planet is much closer to the asteroid belt. Major obliteration events have occurred on Mars as recently as 200 million to 450 million years ago, whereas most of the scarred lunar terrain was produced billions of years earlier.

Recognizable impact craters on Earth range in age from a few thousand to nearly 2 billion years old. For the past 3 billion years, Earth's cratering rate

Figure 105 Saturn's
moon Mimas has a
heavily and uniformly
cratered surface.

(Photo courtesy NASA)

Figure 105 Saturn's moon Mimas has a heavily and uniformly cratered surface.

(Photo courtesy NASA)

has been relatively constant. A major impact has produced a crater 30 miles or more in diameter every 50 to 100 million years. As many as three large meteorite impacts with craters at least 10 miles wide are expected every million years. An asteroid or comet half a mile in diameter with an impact energy of 1 million megatons of TNT, capable of killing one-quarter of the human population, could strike Earth perhaps every 100,000 years or so.

IMPACT CRATERING

When a large meteorite slams into Earth, it expels massive quantities of sediment, propelling debris at 50 times the speed of sound while vaporizing solid rock. In the process, it gouges out a deep crater from several miles to a hundred miles or more across. The extreme pressures and temperatures of the impact destroy virtually all remnants of the meteorite. Therefore, impact cratering is a unique geologic process, with massive amounts of energy released in a small area in a very short interval. Major impact events greatly affected the course of geologic and biologic history.

Many geologic features originally thought to be formed by other geologic forces such as uplift or volcanism (Fig. 107) are now thought to be impact craters. For example, Upheaval Dome near the confluence of the Colorado and Green Rivers in Canyon Lands National Park, Utah, was originally thought to be a salt plug that heaved the overlying strata upward into a huge bubblelike fold 3 miles wide and 1,500 feet high. An alternate interpretation holds that the structure is actually a deeply eroded astrobleme, the remnant of an ancient impact structure gouged out by a large cosmic body striking Earth such as an asteroid or comet.

The magnitude of the energy liberated during impact depends mainly on the speed and size of the impacting body. Asteroids and comets strike Earth at velocities of about 15 miles per second, or twice the speed of the fastest rocket. Generally, comets hit the planet faster than asteroids because their relative velocities are higher. A body with a mass of more than 1,000 tons shoots through the atmosphere practically unhindered by air resistance,

Figure 106 *The Mars* Pathfinder *landing site in* Ares Vallis.

(Photo courtesy NASA)

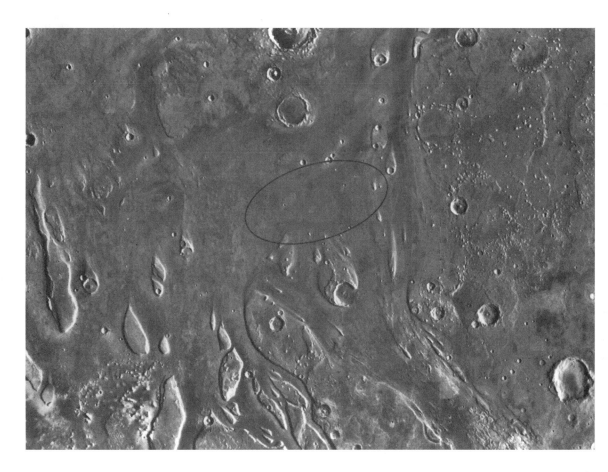

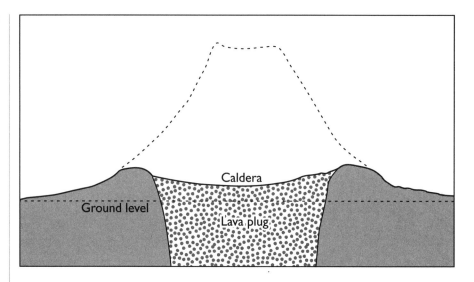

whereas one with a mass of less than 100 tons decelerates to about half its original velocity and often breaks up on entry.

While traveling at high velocities, meteorites transfer to the ground a considerable amount of kinetic energy that is converted to pressure and heat. The larger meteorites generate sufficient pressure and temperature on impact to melt and vaporize the impacting body as well as the rock being impacted. The destruction is so complete that fragments of the impactor are rarely found near large meteorite craters.

The impact lofts the finer material high into the atmosphere. The coarse debris falls back around the perimeter of the crater, forming a high, steep-banked rim (Fig. 108). The rocks are shattered near the impact. The shock

Figure 108 *The structure of a large meteorite crater.*

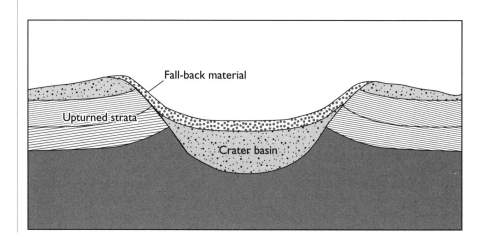

wave passing through the ground produces shock metamorphism in the surrounding rocks, significantly altering their composition and crystal structure. The recognition of shock metamorphism effects has led to the discovery of many craters that would otherwise have gone completely unnoticed.

The discoveries suggest that massive meteorite bombardments during the early part of Earth's history played a major role in shaping the surface of the planet. Most impact craters have disappeared long ago, wiped off the surface of the planet by the same geologic forces that erode the tallest mountains and carve out the deepest ravines. Many craters that do survive this onslaught are simply inaccessible, well hidden beneath the deep sea.

SHOCK EFFECTS

Impact cratering is the only geologic process that produces shock–metamorphic effects. Even the most powerful volcanoes cannot duplicate the blast effects of large meteorite impacts. Rocks also change significantly when subjected to the intense temperatures and pressures during periods of mountain building. Although orogenic processes develop similar temperatures as those of meteorite impacts, the pressures are significantly less.

Temperatures and pressures deep within Earth's crust can equal those required for shock metamorphism. However, they are applied over periods of millions of years, not almost instantaneously as with a meteorite impact. A meteorite producing a crater several miles wide would generate shock compression lasting only a small fraction of a second. Similar conditions have been reproduced during high-yield nuclear weapons testing (Fig. 109).

The most easily recognizable shock effect is the fracturing of rocks into distinct conical and striated patterns called shatter cones. They form most readily in fine-grained rocks with little internal structure, such as limestone and quartzite, a hard metamorphosed quartz sandstone. They develop best at pressures generated by large impactors and provide strong evidence for the existence of meteorite craters that have long since disappeared.

Large meteorite impacts also produce shocked quartz grains with prominent striations across crystal faces (Fig. 110). Abundant minerals such as quartz and feldspar develop these features when high-pressure shock waves exert shearing forces on crystals, producing parallel fracture planes called lamellae. The number of planes and their orientation, which depend on the crystalline structure, is in direct proportion to the shock pressure. At certain shock pressures, a mineral can lose its crystalline structure entirely.

Shock metamorphism can also produce certain high-pressure minerals such as diamonds from carbon and stishovite from quartz. Stishovite is perhaps the best indicator of shock metamorphism by a large meteorite impact.

Figure 109   *A large crater produced by an underground nuclear explosion 300 feet beneath the surface.*

(Photo courtesy U.S. Department of Energy)

Although stishovite can be created within Earth's mantle at depths approaching 400 miles, it reverts into stable quartz before reaching the surface as pressure is relieved. Therefore, any stishovite found above ground had to be formed there, and the only phenomenon known to generate the required pressure is a meteorite impact.

Figure 110   *Lamellae across crystal faces produced by high-pressure shock waves from a large meteorite impact.*

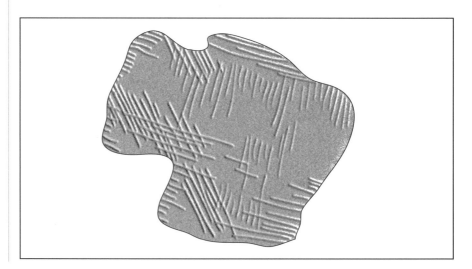

The extremely high pressures of an impact cause minerals such as feldspar and quartz to lose their crystalline structure. These pressures transform the minerals into diaplectic glasses, from the Greek *diaplesso* meaning "to destroy by striking." Unlike conventional glasses, these have an overall shape and composition of the original crystal but have no apparent order or structure.

The impact force also generates extremely high temperatures that fuse sediment into small glassy spherules, which are tiny spherical bodies that resemble volcanic glass. The spherules are similar to the glassy chondrules (rounded granules) in carbonaceous chondrites, which are carbon-rich meteorites, and in lunar soils. Extensive deposits of 3.5-billion-year-old spherules in South Africa are more than 1 foot thick in places. Spherule layers up to 3 feet thick have been found in the Gulf of Mexico and are related to the 65-million-year-old Chicxulub impact structure off Yucatán, Mexico. The melt rocks at the Manicouagan impact structure in Quebec, Canada, were first mistakenly identified as having a volcanic origin.

Impact melt rocks differ from volcanic rocks by originating in nonvolcanic terrain, by containing large amounts of unmelted fragments from the local bedrock, and by having a chemical composition unlike that of any volcanic rocks. Generally, impact rocks are a mixture of various crustal rocks that have been shocked above their melting point. In contrast, volcanic rocks have compositions determined by the melting of selected minerals found beneath Earth's crust.

Certain impact melt rocks contain unusual amounts of trace elements that could only have originated from a meteorite. For instance, the concentration of nickel in an impact melt rock is as much as 20 times greater than the local bedrock. Other elements such as iridium, platinum, osmium, and cobalt that are extremely rare in Earth's crust are more concentrated in meteorites. Therefore, a high concentration of these elements in a melt rock is an excellent indicator of a meteorite impact.

CRATER FORMATION

The formation of impact craters is remarkably similar to the effect of a high-powered rifle bullet fired into the ground. Large meteorites traveling at high velocities completely disintegrate upon impact. In the process, they create craters generally 20 times wider than the meteorites themselves. The crater diameter varies with the type of rock being impacted on due to the relative differences in rock strength. A crater made in crystalline rock can be twice as large as one made in sedimentary rock. The largest impacts not only dig deep craters but raise peculiar mountain chains encircling the impact sites that have since been eroded down to faint rings.

When a meteorite strikes the surface (Fig. 111), it sends a shock wave with pressures of millions of atmospheres down into the rock and back through the meteorite. The shock wave compresses terrestrial rocks and propels them downward and outward from the point of impact, accelerating the target rocks to speeds of several miles per second. As the meteorite burrows into the ground, it forces the rock aside and flattens itself in the process. It is then deflected, and its shattered remains shoot out of the crater. This is followed by spray of shock-melted meteorite and melted and vaporized rock expelled at a high velocity.

As the meteoritic material continues rising through the atmosphere, it forms a rapidly expanding dust plume. The plume grows to several thousand feet across at the base, while the top extends several miles high into the upper atmosphere. Most of the surrounding atmosphere is blown away by the tremendous shock wave generated by the meteorite impact. The giant plume develops into an enormous black dust cloud that punches through the atmosphere similar to the mushroom cloud of a nuclear bomb explosion (Fig. 112). Indeed, many striking similarities exist between the effects of nuclear detonations and large meteorite impacts, except the latter generate no radioactive fallout. Each impact event might have lasted only a few seconds. However, the effects would have been global, prevailing for months or even years.

Figure 111 *The formation of a large meteorite crater.*

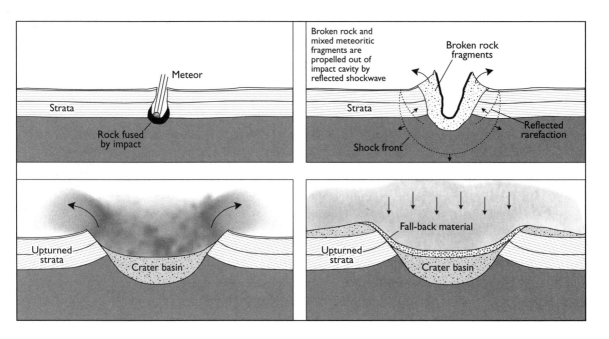

IMPACT STRUCTURES

Many conspicuously circular features have been located on Earth's surface that appear to be impact structures. However, because of their low profiles and subtle stratigraphies along with a covering of dense vegetation, they were previously unrecognized as meteorite craters. Many impact structures appear as large circular patterns in the crust created by the sudden shock of a massive meteorite landing onto the surface.

Some impacts form distinctive craters, whereas others might show only subtle outlines of former impact sites. The only evidence of their existence might be a circular disturbed area with rocks severely altered by shock metamorphism. This process requires the instantaneous application of high temperatures and pressures equal to those found deep in Earth's interior.

Roughly 150 impact craters are known throughout the world (Table 7). More are discovered every year, most of which are within 200 million years

Figure 112 *Detonation of shot Grable atomic bomb in Operation Upshot-Knothole on May 25, 1953.*

(Photo courtesy U.S. Navy)

TABLE 7 LOCATION OF MAJOR METEORITE CRATERS AND IMPACT STRUCTURES

Name	Location	Diameter (in feet)
Al Umchaimin	Iraq	10,500
Amak	Aleutian Islands	200
Amguid	Sahara Desert	
Aouelloul	Western Sahara Desert	825
Baghdad	Iraq	650
Boxhole	Central Australia	500
Brent	Ontario, Canada	12,000
Campo del Cielo	Argentina	200
Chubb	Ungava, Canada	11,000
Crooked Creek	Missouri, USA	
Dalgaranga	Western Australia	250
Deep Bay	Saskatchewan, Canada	45,000
Dzioua	Sahara Desert	
Duckwater	Nevada, USA	250
Flynn Creek	Tennessee, USA	10,000
Gulf of St. Lawrence	Canada	
Hagensfjord	Greenland	
Haviland	Kansas, USA	60
Henbury	Central Australia	650
Holleford	Ontario, Canada	8,000
Kaalijarv	Estonia, USSR	300
Kentland Dome	Indiana, USA	3,000
Kofels	Austria	13,000
Lake Bosumtwi	Ghana	33,000
Manicouagan Reservoir	Quebec, Canada	200,000
Merewether	Labrador, Canada	500
Meteor Crater	Arizona, USA	4,000
Montagne Noire	France	
Mount Doreen	Central Australia	2,000
Murgab	Tadjikistan, USSR	250
New Quebec	Quebec, Canada	11,0000

(continues)

TABLE 7   (CONTINUED)

Name	Location	Diameter (in feet)
Nordlinger Ries	Germany	82,500
Odessa	Texas, USA	500
Pretoria Saltpan	South Africa	3,000
Serpent Mound	Ohio, USA	21,000
Sierra Madera	Texas, USA	6,500
Sikhote-Alin	Sibera, USSR	100
Steinheim	Germany	8,250
Talemzane	Algeria	6,000
Tenoumer	Western Sahara Desert	6,000
Vredefort	South Africa	130,000
Wells Creek	Tennessee, USA	16,000
Wolf Creek	Western Australia	3,000

of age. Thus far, only 10 percent of the expected number of large craters more than 10 miles wide and less than 100 million years old have been found. Although the cratering rate has been somewhat constant during the past few billion years, older craters are less abundant because erosion, sedimentation, and other geologic processes destroyed them.

Many impacts are multiple hits, leaving a chain of two or more craters close together often produced when an asteroid or comet broke up in outer space or upon entering the atmosphere. A rapid-fire impact from a broken-up object 1 mile wide apparently produced a string of three 7.5-mile-wide impact craters in the Sahara Desert of northern Chad. Two sets of twin craters, Kara and Ust-Kara in Russia and Gusev and Kamensk near the northern shore of the Black Sea, formed simultaneously only a few tens of miles apart.

Splattered in almost a straight line across southern Illinois, Missouri, and eastern Kansas are eight large, gently sloping depressions, 2 to 10 miles wide and averaging 60 miles apart. A chain of 10 oblong craters ranging up to 2.5 miles long and 1 mile wide, running along a 30-mile line near Rio Cuarto, Argentina, suggests a meteorite 500 feet wide hit at a shallow angle. It subsequently broke into pieces that ricocheted and gouged their way across the landscape roughly 2,000 years ago.

Some two-thirds of the known impact craters are located in stable regions known as cratons, composed of strong rocks in the interiors of the continents (Fig. 113). The cratons experience low rates of erosion and other

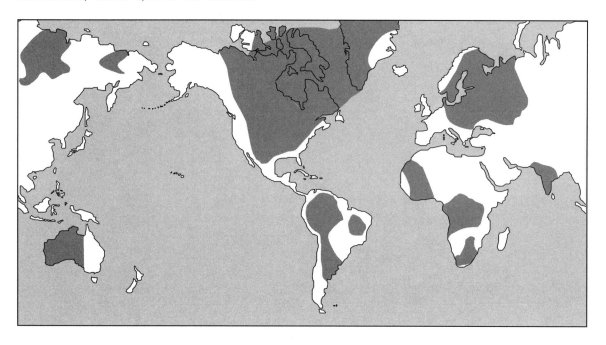

Figure 113   The worldwide distribution of stable cratons.

destructive geologic processes, thereby preserving craters for long periods. Most craters have been found on the cratons of North America, Eurasia, and Australia simply because more exploration has been conducted on those continents than in South America or Africa. Furthermore, since Earth is 70 percent ocean, most meteorites land on the seabed, a vast unexplored terrain.

The two basic forms of meteorite craters are simple and complex, depending on the size and type of impactor. Simple craters such as Arizona's Meteor Crater (Fig. 114) form deep basins and range up to 2.5 miles in diameter. They are often difficult to identify by geology alone because much of the evidence of shock metamorphism lies buried deep beneath the structures. Therefore, many obvious circular features up to a few miles in diameter are hard to prove as having impact origin. Occasionally, as with Meteor Crater, pieces of the impactor lying scattered around the crater readily identify it as having been formed by a large meteorite rather than as the result of volcanic activity.

Larger structures called complex craters are much shallower, up to 100 times wider than their depth. Large craters form in two steps. The impact first creates a deep transient crater. Then, the walls of the initial crater collapse within minutes of the impact, partially filling the transient bowl and enlarging the crater to its final size.

Large craters generally have an uplifted structure in the center, where shocked rocks are exposed, similar to the central peaks observed within craters on the Moon (Fig. 115). The central peak, which takes up about one-tenth of the crater's diameter, forms when rocks in the center rebound upward and lift the crater floor, similar to the way a drop of water splashes up when falling into a pool. For example, the uplifted area inside the 60-mile-wide Manicouagan structure in Quebec is estimated to be about 6 miles wide and resulted from having one-quarter of the thickness of the continental crust raised to the surface.

Surrounding the central peak is an annular trough and a fractured rim. These features are often quite difficult to identify on older craters because they have been severely eroded, forming only vague circular patterns in the terrain. Between the central peak and the outside rim are various materials transformed by the impact. These include melted and fragmented rocks and shocked minerals that help verify the validity of the crater.

Even a structure that looks like an impact crater might have formed instead by other geologic processes, such as volcanism, that produce nearly identical rock formations. The impact origin of complex craters is much easier to confirm than that of simple craters because the uplifted region in the complex crater's center contains abundant shocked rocks. Yet because complex

Figure 114 *Meteor Crater, Coconino County, Arizona.*

(Photo by W. B. Hamilton, courtesy USGS)

Figure 115 *A large meteorite crater on the lunar surface.*

(Photo courtesy NASA)

craters are so shallow and heavily eroded, their uplifted central features are often mistaken for other types of geologic formations. Even the largest and oldest impact structures such as Sudbury in Canada and Vredefort in South Africa, whose original diameters are estimated at 100 miles or more, have been heavily modified over the last 2 billion years, often masking their true identity.

STREWN FIELDS

Tektites (Fig. 116), from the Greek word *tektos* meaning "molten," are glassy bodies formed from the melt of a large meteorite impact. More than half the rocks ejected by an impact can remain molten in the rising plume. After cooling while flying through the air, they fall back to Earth as tektites. Massive meteorite impacts dump millions of tons of tektites stretching over vast areas. Much of the high-flying material is known to deposit halfway around the globe from the site of the impact.

Tektites range in color from bottle green to yellow brown to black. The Cro-Magnon, ancient human ancestors, once prized them as ornaments. Tektites are usually small, about pebble sized, although a few have been known to be as large as cobblestones. Tektites display a variety of shapes from irregular to spherical, including ellipsoidal, barrel, pear, dumbbell, or button shaped. They also have distinctive surface markings that apparently formed while solidifying during their flight through the air.

Tektites differ chemically from meteorites. They have a composition similar to the volcanic glass obsidian but contain much less gas and water. They are much drier than terrestrial rocks and are deficient in elements such as lead, thallium, copper, and zinc, whose compounds become volatile at 1,000 degrees Celsius. Furthermore, they are not composed of microcrystals, a characteristic that is unknown for any type of volcanic glass. Tektites comprise abundant silica similar to the pure quartz sands used to manufacture glass. Indeed, tektites appear to be natural glasses formed by the intense heat generated by a large meteorite impact. The impact flings molten material far and wide. While airborne, the liquid drops of rock solidify into various shapes.

Tektites do not appear to have originated outside Earth because their composition more closely matches that of terrestrial rocks. Their worldwide distribution suggests they were launched at very high velocities by a power-

Figure 116 *A North American tektite found in Texas in November 1985, showing surface erosional and corrosional features.*

(Photo by E. C. T. Chao, courtesy USGS)

Figure 117 *The distribution of tektites in major strewn fields around the world.*

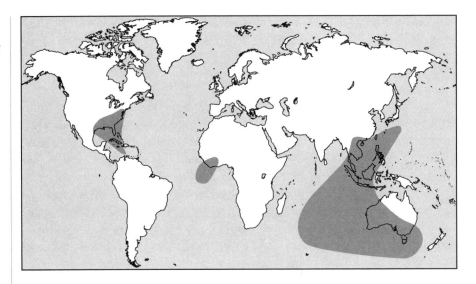

Figure 117 *The distribution of tektites in major strewn fields around the world.*

ful mechanism, such as a massive volcanic eruption or a large meteorite impact. However, terrestrial volcanoes appear to be much too weak to produce the observed strewn fields of tektites that travel nearly halfway around the world.

Impact craters have been associated with known tektite-strewn fields. Tektites have primarily been found littering three extensive strewn fields both on land and on the ocean floor (Fig. 117). The oldest known deposit is the North American strewn field, stretching from the eastern and southern portions of the United States across the Gulf of Mexico and the Caribbean Sea. It dates from about 35 million years ago and might have contained up to 1 billion tons of tektites, much of which has long since eroded away. The Ivory Coast strewn field, extending from the Ivory Coast of Africa into the South Atlantic, is about 1 million years old and contains perhaps 10 million tons of tektites.

Traces of ancient impact structures are also found in seafloor sediments and within the stratigraphic column. The largest trail of tektites, called the Australasian strewn field, stretches from the Indian Ocean, to southern China, southeastern Asia, Indonesia, the Philippines, and Australia. It is around 750,000 years old and comprises perhaps 100 million tons of tektites. The Australasian tektites called javanites are roundish or chunky objects, often shaped like teardrops, dumbbells, or disks, that show little evidence of internal strain. Interestingly, a specimen of one of these tektites was given to Charles Darwin during his sample-gathering voyage around the world in the early 1830s.

Mysterious glass fragments strewn over Egypt's western desert appear to be melts from a huge impact that occurred about 30 million years ago. Large, fist-sized, clear-glass fragments found scattered across the Libyan Desert were analyzed for trace elements, indicating the glasses were produced by an impact into the desert sands. The 2-billion-year-old Vredefort structure in South Africa has been identified as an impact structure based solely on its impact melt and high iridium content.

Shocked metamorphic minerals believed to have originated from a massive impact 65 million years ago when the dinosaurs became extinct are strewn across western North America from Mexico to Canada. Shocked quartz and feldspar grains found in the Raton Basin of northeast New Mexico indicate the impact occurred on land or on the continental shelf a short distance from shore because these minerals are abundant in continental crust and rarely found in oceanic crust.

The large size of the impact grains also suggests that the impact was comparatively close, perhaps within 1,000 miles of the strewn field. The most promising site for the impact crater is the Chicxulub structure in northern Yucatán. It measures between 110 and 185 miles wide, making it one of the largest known craters. An oblique impact from an object arriving from the southeast would send a spray of shocked and molten rock in the direction of the strewn field.

CRATER EROSION

Of all the known impact craters distributed around the world, most are younger than 200 million years. The older craters are less prevalent simply because erosion and sedimentation have destroyed them. Volcanism, folding, faulting, mountain building, weathering, and glaciation have erased or modified most of Earth's ancient history. In addition, these active erosional processes have long eradicated most meteorite impacts. Geologic erosion includes the action of rain, wind, ice, freezing and thawing, and plant and animal activity. The exceptions are craters in the desert and Arctic regions, which experience low rates of erosion.

One interesting impact structure called the Haughton Crater, named for an Arctic geologist, pockmarks Devon Island in the Canadian Arctic (Fig. 118). It was created when a large meteorite slammed into the island at 40,000 miles per hour and excavated a deep crater 15 miles wide some 23 million years ago. Even after millions of years, the crater remains surprisingly intact because little erosion occurs in the desolate tundra.

The Moon and Mars retain most of their craters (Fig. 119) simply because they lack a hydrologic cycle, which has wiped out most impact structures on Earth. The forces of erosion have leveled the tallest mountains, gouged out the deepest canyons, and obliterated all other geologic structures over time. No wonder impact craters cannot escape these powerful weather-

Figure 118 The
location of Haughton
Crater on Devon Island
in the Canadian Arctic.

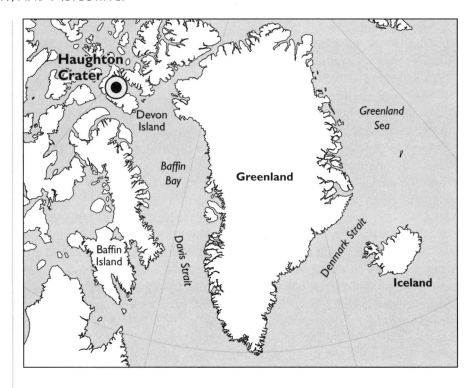

Figure 119 *A heavily*
cratered region on Mars
shows the effects of wind
erosion.

(Photo courtesy NASA)

ing agents. The few exceptions are craters on the ocean floor—which do not experience weathering, craters in the deserts—which do not receive significant rainfall, and craters in the frozen tundra regions—which have remained essentially unchanged for ages.

Craters over 12 miles in diameter and more than 2.5 miles deep appear to be virtually impervious to erosion. They escape destruction because Earth's crust literally floats on a dense, fluid mantle. The process of erosion as material is gradually removed from the continents is delicately balanced by the forces of buoyancy, whereby the crust floats on a sea of molten magma, called isostasy (Fig. 120) from Greek meaning "equal standing." Therefore, erosion can only shave off the upper 2 to 3 miles of the continental crust before its mean height falls below sea level, at which time erosion ceases and sedimentation begins. Massive craters are usually sufficiently deep so that even if the entire continent were worn down by erosion, faint remnants would remain.

Many geophysical methods are employed to detect large impact craters invisible from the air. Seismic surveys could be used to recognize circular distortions in the crust lying beneath thick layers of sediment. The disturbed

Figure 120   The principle of isostasy. Land covered with ice readjusts to the added weight like a loaded freighter. When the ice melts, the land is buoyed upward as the weight lessens.

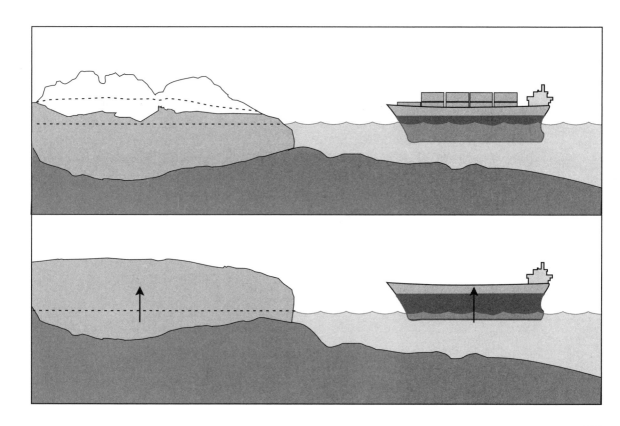

rock usually produces gravity anomalies that could be detected with gravimeters. The fact that many meteorite falls are of the iron-nickel variety suggests they could be detected by using sensitive magnetic instruments called magnetometers.

The surface geology also indicates areas where rocks were disturbed by the force of the impact or outcrop to form a large circular structure. The 10-mile-wide Wells Creek structure in Tennessee lies in an area of essentially flat-lying Paleozoic rocks that were uplifted to form two concentric synclines (downfolded strata) separated by an anticline (an upfolded stratum).

The 65-million-year-old Chicxulub structure, which lies well hidden beneath nearly 1 mile of sediments off the tip of the Yucatán peninsula, was discovered by geophysical methods alone. A seismic survey revealed that the crater has multiple concentric rings, a structure often seen on the Moon and Venus but not on Earth. The crater has a raised inner ring about 50 miles in diameter, another ring about 85 miles across, and an outer ring about 120 miles wide. Signs of an even larger ring also exist, which would make Chicxulub one of the largest impact formations on Earth. Cuttings from oil exploration wells drilled in the area showed signs of impact shock. Ocean core samples taken off the east coast of Florida contained a brownish clay thought to be the "fireball layer," from the vaporized remains of the asteroid itself. In addition, impact debris, including tektites and rubble washed onshore, was found scattered around the region.

A gravity survey also uncovered a distinct bull's-eye pattern of gravity anomalies. The buried crater is outlined by an unusual concentration of sinkholes concentric with the gravity rings. What is unique is the well-defined circular pattern in which the sinkholes were arranged. The impact formed a circular fracture system that acts as an underground aquifer for groundwater, a critical resource since no surface streams are present. The cavity formation in the sinkholes extends to a depth of about 1,000 feet. Its permeability allows the ring to act as a conduit carrying water out to sea.

After discussing the formation of meteorite craters, the next chapter examines the effects to Earth from large asteroid and comet impacts.

8

IMPACT EFFECTS
THE GLOBAL CHANGES

This chapter examines the worldwide effects of large asteroid or comet collisions with Earth. Major global changes would result from the impact of a large asteroid or comet. The collision would set the planet ringing like a giant bell. The jarring action would produce powerful earthquakes and violent volcanic eruptions. An impact into the ocean would send gigantic tsunamis racing toward nearby shores. The impact could also reverse the planet's magnetic field, and some reversals in the past have been strongly linked to meteorite craters.

The impact would loft a thick blanket of dust into the atmosphere, shutting out the Sun and chilling the planet. The cold and dark would halt photosynthesis, knocking the bottom out of the food chain and inflicting mass starvation. The impact could also cause a slight shift in Earth's spin axis, resulting in glaciation. So much damage would beset Earth that extinction of species would surely follow.

GLOBAL EFFECTS

Massive dust storms are common on Mars (Fig. 121). When *Mariner 9* entered orbit around Mars in 1971, the spacecraft encountered a global dust storm that

Figure 121 *A Martian dust storm observed by* Mariner 9 *continued for months before clearing enough to view the surface.*

(Photo courtesy NASA)

obscured the planet from view. The dust storm continued for months before finally clearing away sufficiently to allow the orbiter to photograph the surface. The scale of the Martian dust storm, unheard of on Earth, prompted speculation on the global effects of thermonuclear war known as "nuclear winter." A remarkably similar effect would be caused by the impact of a large asteroid or comet.

The environmental effects of a major impact have been compared with the 1883 eruption of Krakatoa in Indonesia, which virtually destroyed the island. The explosion was heard 3,000 miles away. Barometers around the world registered the event as the shock wave circled the globe three times. The blast sent a 100-foot tsunami crashing down on nearby shores, drowning 36,000 people. Tidal gauges as far away as Europe detected the giant tsunamis generated by the force of the eruption. Volcanic dust rose 50 miles into the atmosphere, where it shaded the planet and brought down global temperatures for several years, seriously disrupting agriculture, which only added to the death toll.

Geologic history can provide clues about the effect on Earth following a collision with a large cosmic body when skies become clogged with dust and smoke. The evidence is hidden in sediments laid down at the end of the Cretaceous period 65 million years ago, when the dinosaurs mysteriously vanished (Fig. 122). Only something akin to global nuclear war could have destroyed life on such a grand scale. Indeed, the environmental consequences

Figure 122   *A dinosaur boneyard at the Howe Ranch quarry near Cloverly, Wyoming.*

(Photo by G. E. Lewis, courtesy USGS)

165

of nuclear war and a large meteorite impact would be strikingly similar except for the lack of significant amounts of radioactivity following a meteorite strike.

If a large celestial body such as an asteroid or comet slams into Earth, the impact would produce thick dust clouds, powerful blast waves, towering tsunamis, extremely toxic gases, and strong acid rains that would be immensely damaging to the environment. Perhaps the worst environmental hazards would be produced by large volumes of suspended sediment in the atmosphere from material blasted out of the crater along with the vaporized impactor.

The impact would pump several hundred million tons of soil and dust particles into the atmosphere along with soot from burning forests. Global atmospheric mixing would enable the dust cloud to encircle the entire planet. Solar radiation would also heat the darkened, sediment-laden layers in the atmosphere and cause a thermal imbalance that would radically alter weather patterns, turning much of the land surface into a barren desert.

Horrendous dust storms driven by maddening winds would rage across whole continents, further clogging the skies. Moreover, an asteroid or comet slamming into the ocean floor would release enough heat to generate massive hurricanes called *hypercanes* that would reach 20 miles high with sediment-laden winds approaching 500 miles per hour. The dust along with soot from continent-sized wildfires clogging the skies would bring darkness at noon.

Roughly one-third of the land area is covered with forests and adjacent brush and grasslands. Heat produced by the compression of the atmosphere and impact friction with molten rock flung in all directions along with glowing bits of impact debris flying past and back through the atmosphere would set global forest fires that produce a thick blanket of smoke. This would be considerably worse than the present conflagration that engulfs the Amazon rain forest (Fig. 123). The blaze would burn perhaps one-quarter of all vegetation on the continents, turning a large part of Earth into a smoldering cinder.

Forest fires also ignited by excessive dry conditions brought on by the lack of evaporative processes would burn totally out of control, destroying hundreds of thousands of square miles of land in a matter of days. The inferno would consume as much as 80 percent of the terrestrial biomass. A heavy blanket of dust and soot would encircle the entire planet and linger for months, cooling the climate and halting photosynthesis. The forest and grassland fires would also generate large amounts of nitrous oxide that would precipitate as strong acid rain. A catastrophe on such a large scale would destroy most terrestrial habitats, causing tragic extinctions.

The dust and smoke injected into the atmosphere would prevent sunlight from reaching the surface, which would severely reduce condensation and precipitation. This action would halt the cleansing effect of rain and allow the dust and smoke to linger for extended periods. The reduction of solar energy on the surface of the ocean would severely restrict evaporation and

Figure 123 *The Amazon basin of South America is obscured by smoke from clearing and burning of the tropical rain forest, viewed from the space shuttle* Discovery *in December 1988.*

(Photo courtesy NASA).

consequently the amount of rainfall, allowing atmospheric particles to remain suspended for an abnormally long time.

The suspended sediment and soot clogging the skies would greatly reduce the level of sunlight reaching the surface and halt photosynthesis, causing plants to wither and die. Vegetation would also suffer from a lack of rainfall. Drought conditions would prevail over large areas of Earth, turning it into a global desert. Furthermore, the reduced sunlight would drop surface temperatures significantly, causing plants to cease growing. This would severely disrupt terrestrial food chains and cause extinctions of massive proportions.

The continual darkness would kill off marine phytoplankton, which require sunlight for photosynthesis. The death of these primary producers

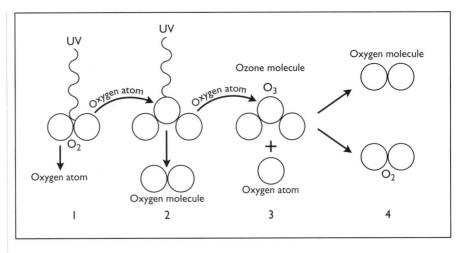

Figure 124 *The life cycle of an ozone molecule. (1) Ultraviolet (UV) radiation splits an oxygen molecule into two oxygen atoms. (2) One of these atoms combines with another oxygen molecule to create an ozone molecule. The ozone molecule traps UV, liberating an oxygen molecule and an oxygen atom, to reform as an ozone molecule. (3) The addition of another oxygen atom creates (4) two new oxygen molecules.*

would produce deadly consequences on up the food chain, resulting in catastrophic extinctions in the ocean as well. In the tropical regions, where the indigenous organisms are particularly sensitive to changes in temperature, vast arrays of marine species would quickly disappear. The ocean would experience a cataclysmic die-off of species on a scale equal to the great extinctions of the geologic past.

A massive bombardment of meteorites would also strip away the upper atmospheric ozone layer, bathing Earth in the Sun's deadly ultraviolet rays, disrupting the regeneration of new ozone molecules (Fig. 124). The increased radiation would kill land plants and animals along with primary producers in the surface waters of the ocean. The destruction of the ozone layer by nitrous oxides generated by intense temperatures created by the impact and the forest fires it sets would enable strong solar ultraviolet radiation to reach Earth's surface after the smoke and dust finally cleared away.

A large dose of ultraviolet radiation is deadly to plants and animals. Therefore, entire terrestrial ecosystems would be destroyed. Agricultural production desperately needed for recovery would fall off sharply, bringing on mass starvation of unprecedented levels. The human species, already teetering on the brink of disaster from overpopulation, would be all but totally wiped out.

TECTONISM

Large meteorite impacts create so much disturbance in Earth's thin crust that volcanoes and earthquakes could become active in zones of weakness, causing additional environmental destruction. Earthquake faults poised to break could be triggered by the powerful seismic activity of a continental impact. Convincing evidence suggests that Earth was struck by a large asteroid or comet nucleus at the end of the Cretaceous. The shock wave created by the force of such an impact would have registered a reading of 13 on the earthquake magnitude scale, 1 million times more powerful than the strongest quakes ever recorded.

Some types of shallow earthquakes with magnitudes greater than 5.0 are more easily triggered by outside events such as meteorite impacts. Even the gravitational attractions of the Sun and the Moon can pull together on Earth to cause some earthquake fault systems to rupture. The energy released by even a moderate earthquake is tremendous (Fig. 125), equivalent to 100 Hiroshima-sized atomic bombs.

Major meteorite impacts can also initiate flood basalt eruptions (Fig. 126 and Table 8) by fracturing the crust or by disturbing mantle convection, causing plumes of hot magma to rise to the surface. Corresponding cycles in the starting dates of major flood basalts, nonmarine extinctions, and impact cratering have occurred over the last 250 million years. However, explosive eruptions would have been more effective at lofting volcanic materials high into the atmosphere than the comparatively mild flood basalts.

Volcanoes erupting by the jolt from the bombardment of one or more large meteorites could inject large quantities of volcanic ash into the atmosphere. Craters of extremely large meteorite impacts could temporarily reach depths of 20 miles or more, exposing the hot mantle below. The uncovering of the mantle in this manner would result in a colossal volcanic explosion, releasing tremendous amounts of ash into the atmosphere that could exceed the atmospheric products generated by the meteorite impact itself.

Perhaps the most convincing evidence of an impact coinciding with a volcanic eruption occurred in India at the end of the Cretaceous period, about 65 million years ago. A giant rift sliced down the west side of India, and huge volumes of molten rock poured onto the subcontinent (Fig. 127). Nearly 500,000 cubic miles of lava up to 8,000 feet thick were released over a period of half a million years, blanketing much of west-central India in an area known as the Deccan Traps, meaning the "southern staircase." It was the largest volcanic eruption in the past 200 million years. Multiple layers of basalt lava, each upward of hundreds of feet thick, covered the ground.

Quartz grains shocked by high pressures possibly generated by a large meteorite impact were found lying just beneath the immense lava flows. The deposits appear to be linked to an apparent meteorite impact in the Amirante Basin 300 miles northeast of Madagascar, near India's location when the subcontinent was drifting toward southern Asia. The flood basalts themselves could have contributed upward of 30,000 tons of iridium to the K-T boundary layer. The eruptions would have dealt a major blow to the climatic and ecological stability of the planet and caused widespread extinctions.

Volcanoes play a direct role in Earth's climate. Large volcanic eruptions spew massive quantities of weather-altering ash and aerosols into the atmosphere,

Figure 125 *A collapsed bridge on Interstate 880 in San Francisco, caused by the October 17, 1989, Loma Prieta Earthquake, Alameda County, California.*

(Photo by G. Plafker, courtesy USGS)

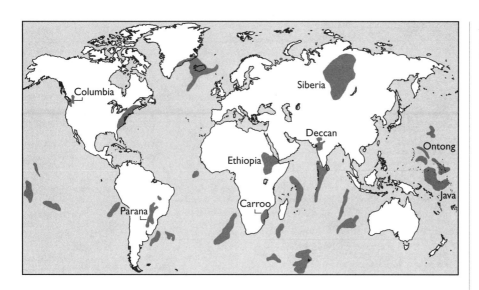

Figure 126 *Areas affected by flood basalt volcanism.*

such as what occurred during the 1980 Mount St. Helens eruption (Fig. 128). These substances block out sunlight and cool the planet. Volcanic dust also absorbs solar radiation, thereby heating the atmosphere internally. This causes thermal imbalances and unstable climate conditions.

Massive quantities of greenhouse gases spewed from volcanoes could cause global warming. In addition, volcanic eruptions produce acid. However,

TABLE 8   FLOOD BASALT VOLCANISM AND MASS EXTINCTIONS

Volcanic episode	Million years ago	Extinction event	Million years ago
Columbian River, USA	17	Low-mid Miocene	14
Ethiopian	35	Upper Eocene	36
Deccan, India	65	Maastrichtian	65
		Cenomanian	91
Rajmahal, India	110	Aptian	110
Southwest African	135	Tithonian	137
Antarctica	170	Bajocian	173
South African	190	Pliensbachian	191
E. North American	200	Rhaectian/Norian	211
Siberian	250	Guadalupian	249

Figure 127 *The Deccan Traps flood basalts in India.*

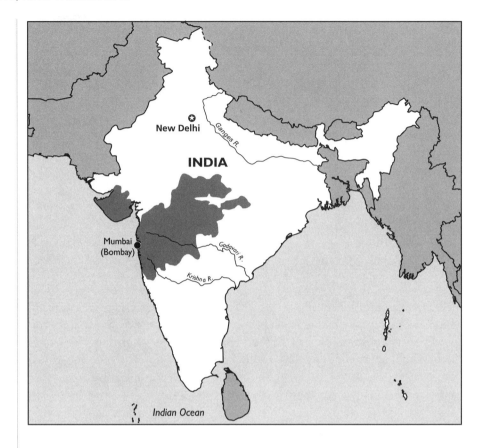

Figure 127 *The Deccan Traps flood basalts in India.*

the amount of acid rain generated by a cometary event would be 1,000 times greater. Massive volcanic eruptions have also been cited as the primary cause for the extinction of the dinosaurs. Therefore, the giant brutes could have been killed off without the aid of a meteorite.

TSUNAMIS

The largest waves generated by fierce storms at sea and tsunamis from large undersea earthquakes or volcanic eruptions rarely exceed more than 100 feet in height. However, none of these waves can compare with the mega-tsunamis generated when a mountain-sized asteroid or comet lands in the ocean. The enormous tsunamis generated by a meteorite splashdown at sea would be particularly hazardous to onshore and nearshore inhabitants much more so than the largest waves created by massive earthquakes, such as those that destroyed large parts of Alaska in 1964 (Fig. 129). The shock of the impact could also send gigantic mudslides crashing down the continental slopes into deeper waters, which would generate tremendous tsunamis.

If an asteroid or comet landed in the ocean, it would instantly evaporate massive quantities of seawater, saturating the atmosphere with billowing clouds of steam. This added burden would dramatically raise the density of the

Figure 128 *A large eruption cloud from the July 22, 1980, eruption of Mount St. Helens.*

(Photo courtesy USGS)

Figure 129 *Tsunami damage at Kodiak from the 1964 Alaskan earthquake.*

(Photo courtesy NOAA)

atmosphere and greatly increase its opacity, making conditions highly difficult for sunlight to penetrate. Furthermore, if the impactor vaporized thick layers of limestone similar to those in the Caribbean, enormous amounts of carbon dioxide injected into the atmosphere could cause a runaway greenhouse effect that would bake the planet.

Upon impact, the asteroid or comet would create a conical-shaped curtain of water as billions of tons of seawater splashed high into the atmosphere. The atmosphere would become oversaturated with water vapor. Thick cloud banks would shroud the planet, cutting off the Sun and turning day into night. The most massive tsunamis ever imagined would race outward from the impact site. Thousand-foot-high waves would traverse completely around the world. When striking seashores, they would travel hundreds of miles inland, devastating everything along the way.

Evidence suggests that an asteroid appears to have landed in the Indian Ocean about 300 miles northeast of Madagascar off the east coast of Africa 65 million years ago. The impact apparently carved out the Amirante Basin circular structure, a nearly 200-mile-wide depression that is mostly intact on the southern edge of the Seychelles Bank. Additional evidence that an asteroid landed in this region is a massive landslide that covers 7,500 square miles on the east coast of Africa. The landslide could have been triggered by a tremendous tsunami generated at the time of the impact.

The most compelling evidence that a large asteroid or comet nucleus landed in the sea is the Chicxulub structure, one of the largest known impact craters on Earth. It lies beneath about 1 mile of sedimentary rock on Mexico's northern coast of the Yucatán peninsula. At the end of the Cretaceous, the impact site was submerged under perhaps 300 feet of water. If the meteorite landed on the seabed just offshore, 65 million years of sedimentation would have long since buried it beneath thick deposits of sand and mud.

Furthermore, rocks in the Caribbean and Gulf of Mexico appear to bear the scars of an impact in the region. A splashdown in the ocean would have created an enormous tsunami that scoured the seafloor and deposited its rubble onto nearby shores. Millions of tons of ocean floor rubble from the impact washing up on shore helped pinpoint the location of the crater.

The tsunami would also have surged up onto the land and dragged rocks and organic debris back into the deep sea. Indeed, fossilized bits of wood were found in exposed marine sediments that were uplifted from a depth of nearly 2,000 feet after being deposited 65 million years ago. The giant swells would have sloshed back and forth across the Gulf of Mexico as though it were a huge bowl of water tilted to-and-fro.

MAGNETIC REVERSALS

The geomagnetic field protects Earth against dangerous cosmic radiation originating from outer space and from the solar wind (Fig. 130). The field reverses in a highly irregular pattern that appears to be a random process, and half the time it regenerates with the opposite polarity. Over the last 170 million years, the geomagnetic field has reversed a total of nearly 300 times. During the last 30 million years, it has reversed on average roughly four times every million years.

The last field reversal was about 780,000 years ago, signifying the planet is well overdue for another one. Earth presently appears to be in a period of gradual magnetic field decline. If this process continues, the weakening of the field could lead to a reversal sometime during the next few thousand years.

Geomagnetic field reversals have occurred during all subsequent geologic periods since the Precambrian era. No evidence suggests that one polarity was

Figure 130 *The magnetosphere shields Earth from cosmic rays in the solar wind.*

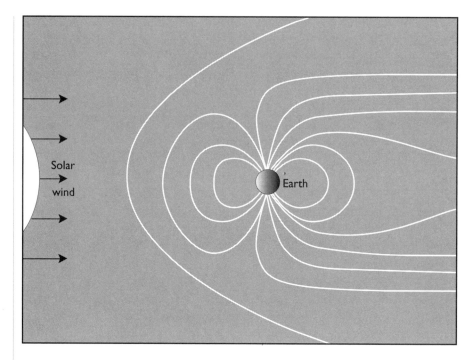

favored over the other for long durations except possibly during the Cretaceous period between 123 million and 83 million years ago, called a magnetic *superchron*. During that time, apparently no reversals occurred for 40 million years. Furthermore, evidence suggests the rate of geomagnetic field reversals peaked during mass extinctions.

The field reversals might occur along with reversals of convective currents in the core (Fig. 131), responsible for generating the magnetic field by the so-called dynamo effect. A dynamo is a device that generates an electrical current by the rotation of a conducting medium through a magnetic field, which in turn reinforces the magnetic field (Fig. 132). It is a very unstable device and can self-reverse polarity in a seemingly chaotic manner. In the case of Earth's dynamo, slight rotation rate differences between the solid inner core and the mantle generate the geomagnetic field, with the liquid outer core acting as an electrical conductor.

Reversals in the convective currents in the fluid outer core could result from fluctuations in the level of turbulence in the core caused by heat loss at the core-mantle boundary and the progressive growth of the solid inner core, which provides the gravitational energy to power the dynamo. Changes in core pressure brought on by tectonic events, glaciation, and large meteorite impacts could also cause fluctuations in the core's turbulence and consequential reversal of the geomagnetic field.

Following a period of several hundred thousand to 1 million or more years of stability, a sudden drop in the magnetic field intensity to about 20 percent of normal occurs in less than 2,000 years, followed by a delay of about 20,000 years before the reversal finally begins. The magnetic field then abruptly collapses, reverses, and slowly builds back to its normal strength. The entire process takes about 10,000 years to complete. Upward of 1,000 years might elapse before the field finally regains its full magnetic intensity. Meanwhile, the excursions of the magnetic poles cause them to wander around the polar regions (Fig. 133).

The geomagnetic field has reversed 11 times during the last 4 million years. About half of the more recent magnetic reversals appear to be associated with large impacts. A correlation between changes in the magnetic field and events taking place on Earth's surface has long been suspected. A comparison of many reversals with known fluctuations in the climate shows a striking agreement. Magnetic reversals occurring roughly 2.0, 1.9, and 0.7 million years ago coincide with unusual cold spells (Table 9).

Moreover, the last two magnetic reversals correspond to the impacts of large asteroids or comet nuclei, which are associated with two tektite-strewn

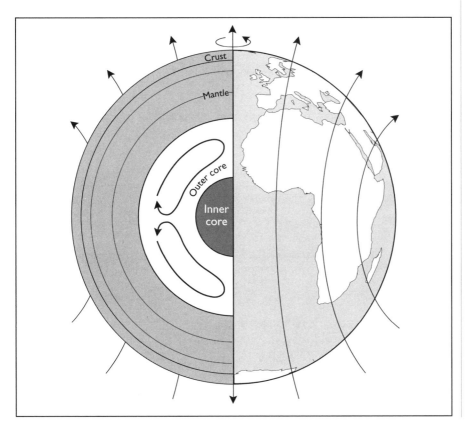

Figure 131 Earth's magnetic field is generated by the slow rotation of the solid, metallic inner core with respect to the liquid outer pole.

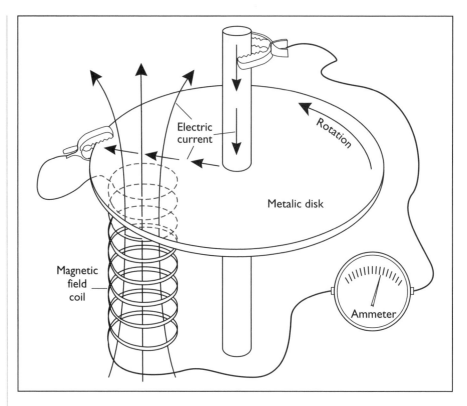

fields. One of these is located in Australasia, whose source crater is on the Asian mainland and dates to about 730,000 years ago. The other is located in the Ivory Coast region, whose source is the Bosumptwi crater in Ghana, West Africa, which dates to about 900,000 years ago and is presently filled with a 6-mile-wide lake.

Perhaps the most striking example of a meteorite impact causing a magnetic reversal is the 15-mile-wide Ries crater in southern Germany, which dates to about 14.8 million years ago. The magnetization of the Ries crater fallback material partially melted by the impact indicates the polarity of the geomagnetic field at the moment of collision. However, the first sediments deposited in the newly formed crater show the opposite polarity, which suggests a reversal must have occurred soon after the impact.

During a geomagnetic field reversal, when Earth lets down its magnetic shield, the climate cools. Variations in the magnetic field intensity over the past several hundred thousand years show a close agreement with variations in surface temperatures. When the field weakens, more cosmic rays can penetrate to the lower atmosphere and warm it, causing thermal imbalances that affect the climate. Furthermore, when the geomagnetic field strength is low, the atmosphere is exposed to the solar wind and high-intensity cosmic radiation. The increased bombardment of air molecules could influence the composition of the upper atmosphere by making more nitrogen oxides, which could block out sunlight and thereby alter the climate.

The climate changes resulting from comet showers could also cause reversals of Earth's magnetic field. The comet strikes would produce debris and smoke from widespread fires, darkening the atmosphere and creating cold climatic conditions. The drop in temperatures would increase glaciation of land areas near the poles. As a result, the accumulation of ice would suddenly drop sea levels.

The shift in Earth's mass due to the accumulation of ice in the polar regions could be sufficient to alter its rotation. The change in rotation speed

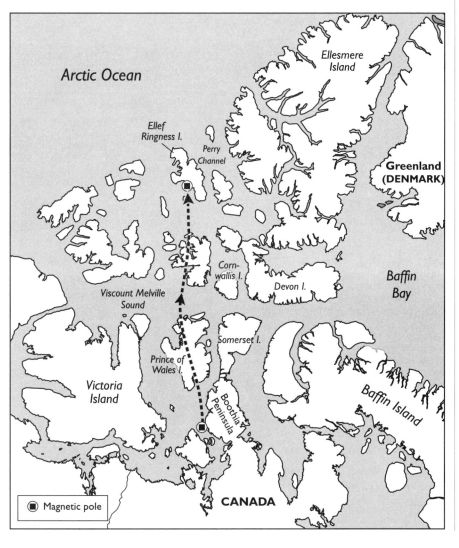

Figure 133 Polar wandering over the last 150 years from Boothia Peninsula to Ellef Ringness Island, covering a distance of about 450 miles in the Canadian Arctic.

TABLE 9   COMPARISON OF MAGNETIC REVERSALS WITH OTHER PHENOMENA (DATES IN MILLIONS OF YEARS)

Magnetic reversal	Unusual cold	Meteorite activity	Sea level drops	Mass extinctions
0.7	0.7	0.7		
1.9	1.9	1.9		
2.0	2.0			
10				11
40			37–20	37
70			70–60	65
130			132–125	137
160			165–140	173

would, in turn, disrupt convection currents in the liquid outer core that produces the magnetic field, causing it to reverse polarity. Not every comet shower would result in a geomagnetic reversal, however, especially during warm periods when they probably would not cool Earth sufficiently to cause glaciation.

GLACIATION

Many periods of glaciation have occurred throughout Earth's history. A single large meteorite impact or a massive meteorite shower would eject so much debris into the global atmosphere that it would block out the Sun for many months or years. This could possibly bring down surface temperatures sufficiently to initiate glaciation. Major changes would occur after the collision of large objects with Earth, which is a frequent occurrence on geologic time scales. Such impacts would create huge clouds of airborne particles, reducing the sunlight reaching the polar oceans, thereby triggering the accumulation of ice.

About 2.3 million years ago, an asteroid or comet apparently impacted on the Pacific seafloor roughly 700 miles westward of the tip of South America (Fig 134). The impact would have sent aloft some 500 billion tons of Sun-shading sediments into the atmosphere. Geologic evidence suggests the climate changed dramatically between 2.2 and 2.5 million years ago, when continental glaciers began to spread over large parts of the Northern Hemisphere (Fig. 135).

The dust and smoke blocking out the Sun would cause a rapid cooling of Earth's surface by 20 degrees Celsius or more. The cooling would persist from several months to more than a year, bringing freezing weather conditions

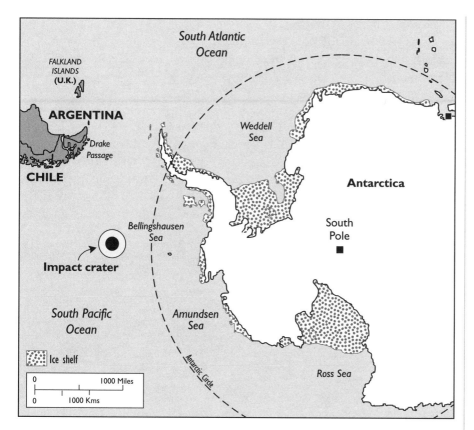

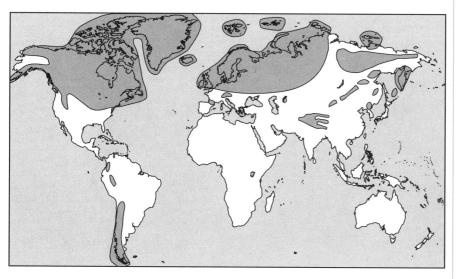

Figure 135   *The extent of Pleistocene glaciation.*

even in the middle of summer. Ice floes in the polar seas would also increase. As a whole, though, the ocean would not freeze over entirely due to its large heat capacity. However, the extreme temperature difference between land and sea would produce violent coastal storms.

A large meteorite impact could alter Earth's orbital motions. This might explain why the temperature of the planet has been steadily dropping since the end of the dinosaur era. The ice ages have been linked to changes in Earth's orbital variation, including the ellipticity of its orbit, the tilt of its spin axis (nutation), and precession of the equinoxes (Fig. 136). Earth's orbital variations on timescales of about 100,000 years over the last 800,000 years coincide with the periodicity of the ice ages. The orbital variations affect the amount of sunlight received from one season to the next, thus altering the contrast between summer and winter temperatures.

Figure 136 *Earth's orbital variations affect the amount of sunlight Earth receives from the Sun. These include the ellipticity of the orbit, the tilt of the spin axis (nutation), and precession of the equinoxes.*

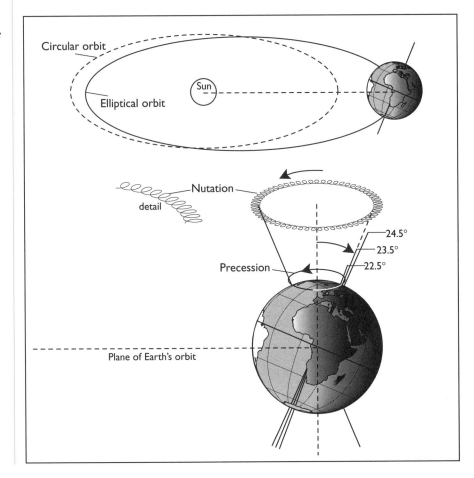

182

A change in the angle of Earth's spin axis would alter the amount of solar radiation impinging on certain latitudes during different times of the year. The stretching of Earth's orbit around the Sun would place the planet several million miles farther away during one season, thereby lowering the input of solar radiation. Although seasonal changes might be extreme, the orbital variations do not significantly reduce the average amount of solar input during an entire year.

Every 100,000 years, the angle of Earth's orbit around the Sun could change in such a way as to bring the planet closer to a region of the solar system that contains large amounts of dust and rocky debris. The cosmic dust entering the atmosphere could trigger long-term global cooling by reflecting the Sun's warming rays back into space directly or indirectly by enhancing cloud formation, thereby reducing the amount of sunlight reaching the ground. Larger chunks of debris could penetrate the atmosphere, releasing dust and gas as they burn up, possibly perturbing the atmosphere sufficiently to start an ice age.

MASS EXTINCTION

A large asteroid crashing down on Earth creates a huge explosion that ejects massive amounts of sediment and excavates a deep crater. The finer material lofts high into the atmosphere, where it shades the planet and lowers global temperatures. In addition, acids produced by a large number of meteors or comets entering the atmosphere could upset the ecological balance by introducing strong acid rains into the environment. A massive bombardment of meteors or comets could also erode the protective ozone layer, allowing the Sun's deadly ultraviolet rays to reach the surface. The increased radiation would kill species on land as well as primary producers in the surface waters of the ocean such as plankton (Fig. 137), which are small floating plants and animals in the sea.

The impacts of large cosmic bodies would result in almost instantaneous extinctions. A large asteroid striking Earth with a force equal to 1,000 eruptions of Mount St. Helens would send aloft some 500 billion tons of sediment into the atmosphere. The impact would also excavate a crater deep enough to expose the molten rocks of the mantle, creating a massive volcanic eruption. Along with huge amounts of dust generated by the impact itself, large quantities of volcanic ash injected into the atmosphere could choke off the Sun.

Impact friction and the compression of the atmosphere would provide sufficient heat to ignite global forest fires. The wildfires would consume one-quarter of the land surface, wiping out land species on a grand scale. The flames would destroy terrestrial habitats and cause extinctions of massive proportions. A heavy blanket of dust and soot would cover the entire globe and linger for months. The atmospheric pollution would cool Earth and halt photosynthesis, killing off species in large numbers.

Figure 137
Coccolithophore were
marine planktonic plants.

Figure 137
Coccolithophore were marine planktonic plants.

A year of darkness under a thick brown smog of nitrogen oxide would ensue. Global rains as corrosive as battery acid would infiltrate into the ground, and runoff would be poisoned by trace metals leached from the soil and rock. Only plants in the form of seeds and roots could survive such an onslaught. The high acidity levels in the ocean would kill off plankton on a terrific scale, knocking out the bottom of the marine food chain and starving out species higher up the food web. Land animals living in burrows and creatures occupying lakes buffered against the acid would have been well protected from the slaughter.

Three or more meteorite impacts have been strongly linked to mass extinctions of species. All impact-related extinctions are distinguished by anomalous concentrations of iridium and deposits containing glassy beads called microtektites that could originate only by collisions with asteroids or comets. Furthermore, the extinctions show an apparent periodicity coincident with the recurrent bombardment of comets.

The first mass extinction event connected to the impact with Earth of one or two large asteroids or comets occurred near the end of the Devonian period about 365 million years ago, and wiped out many tropical marine groups. Evidence for the meteorite impacts is supported by the discovery of microtektites in Belgium and the Hunan province of China that were possibly related to the Siljan crater in Sweden and a series of craters in the Sahara Desert of northern Chad. The deposits also include an anomalous iridium content, which suggests an extraterrestrial source.

The greatest extinction on Earth occurred at the end of the Permian period 250 million years ago, which eliminated more than 95 percent of all species, mostly marine species such as the bryozoan (Fig. 138). Some 75 percent of the amphibian families and more than 80 percent of the reptilian families disappeared along with the vast majority of marine invertebrates. The fossil record suggests the great extinction arrived in two waves spaced 5 million years apart. About 70 percent of species disappeared during the first event, and 80 percent of those remaining died out in the second episode. Only a major environmental catastrophe such as an asteroid or comet impact or a huge volcanic eruption could have caused biological havoc on this scale.

A giant meteorite apparently slammed into Earth at the end of the Triassic period about 210 million years ago. It created a huge impact structure in Quebec, Canada, outlined by the 60-mile-wide Manicouagan reservoir. The gigantic explosion appears to have coincided with a mass extinction over a short period, geologically speaking, that killed off 20 percent or more of all families of animals, including nearly half the reptile families. The extinction forever changed the character of life on Earth and led to the evolution of the direct ancestors of all modern animals. As a direct result of the meteorite impact, the dinosaurs grew from comparably small creatures to huge giants as indicated by their large fossilized bones and impressive trackways (Fig. 139).

The most celebrated extinction event resulted in the death of the dinosaurs and three-quarters of all other species at the end of the Cretaceous period 65 million years ago. Two or perhaps three asteroids or comets might have struck

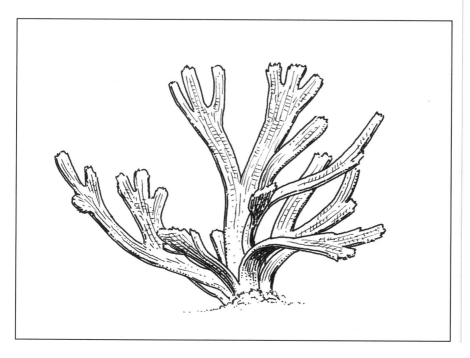

Figure 138 *The extinct bryozoans were major Paleozoic reef builders.*

Earth and left a legacy that has not been duplicated at any other time in geologic history. Rocks defining the end of the period contain the strongest iridium anomalies and other chemical evidence pointing to meteorite impacts. Therefore, the dinosaurs could have been both created and destroyed by asteroids.

The second strongest link between meteorite impacts and mass extinctions occurred when two or more large meteorites landed on Earth

Figure 139 *A dinosaur trackway in Tarapaca Province, Chile.*

(Photo by R. J. Dingman, courtesy USGS)

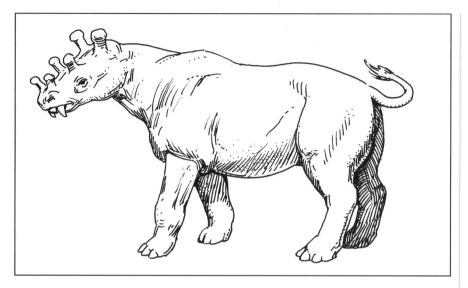

Figure 140 *An extinct Eocene five-horned, saber-toothed, plant-eating mammal.*

near the end of the Eocene epoch around 37 million years ago. Concentrations of microtektites and anomalously large amounts of iridium indicative of meteorite impacts were found in sediments throughout the world. The Eocene extinctions in the sea were spread over a period of 2 to 3 million years, possibly following a full-blown comet shower. Moreover, several impact craters have been dated to this time. The impacts also correlate with the disappearance of the archaic mammals (Fig. 140). These were large, grotesque-looking animals that were probably overspecialized and therefore unable to adapt to the abruptly changing climatic conditions possibly caused by cosmic invaders.

After discussing the effects of asteroid and comet collisions on Earth, the next chapter examines an apparent periodicity of extinctions of species resulting from meteorite impacts.

9

DEATH STAR

IMPACT EXTINCTION OF SPECIES

This chapter examines the apparent periodicity of mass extinctions by meteorite impacts. The disappearance of large numbers of species at certain junctures in Earth's history (Table 10) appears to have both celestial and terrestrial causes. The geologic time scale also implies the mass deaths might be periodic as though some great natural clock controlled the rate of extinctions. Over the last 250 million years, 10 extinction events have been noted that show an apparent periodicity.

The extinction cycle might be attributed to cosmic phenomena, such as Earth's movement through the galactic midplane, where gravity disturbances might break loose Earth-bombing comets from the Oort cloud. A companion brown dwarf star that passes near the Oort cloud possibly orbits the Sun, or a mysterious 10th planet might regularly swing in and out of the Kuiper comet belt. The encounter could disturb a large number of comets and hurl them toward the inner solar system, some of which might rain down onto Earth.

SUPERNOVAS

During the Sun's journey around the center of the Milky Way galaxy, completing the circuit about every 250 million years, it oscillates up and down

TABLE 10   RADIATION AND EXTINCTION OF SPECIES

Organism	Radiation	Extinction
Mammals	Paleocene	Pleistocene
Reptiles	Permian	Upper Cretaceous
Amphibians	Pennsylvanian	Permian-Triassic
Insects	Upper Paleozoic	
Land plants	Devonian	Permian
Fish	Devonian	Pennsylvanian
Crinoids	Ordovician	Upper Permian
Trilobites	Cambrian	Carboniferous & Permian
Ammonoids	Devonian	Upper Cretaceous
Nautiloids	Ordovician	Mississippian
Brachiopods	Ordovician	Devonian & Carboniferous
Graptolites	Ordovician	Silurian & Devonian
Foraminiferans	Silurian	Permian & Triassic
Marine invertebrates	Lower Paleozoic	Permian

perpendicular to the galactic plane. The Sun crosses the plane of the galaxy about every 32 million years, coinciding with one of the major extinction cycles. The apparent regularity of the cometary extinctions might therefore be attributed to Earth's movement through the galactic midplane. Thick gas and dust clouds could produce gravity anomalies sufficiently strong to break loose Oort cloud comets and launch them toward Earth.

The Oort cloud is a sphere of comets surrounding the Sun at a distance of about 100,000 astronomical units, or AU, the average distance between Earth and the Sun, or 93 million miles. It is a vast reservoir of comets with several trillion bodies, having an aggregate mass of 40 Earths. The random gravitational jostling of stars passing nearby knocks some Oort cloud comets out of their stable orbits and deflects their paths toward the Sun. About a dozen stars pass within 200,000 AU of the Sun every million years. These close encounters are sufficient to stir the cometary orbits, sending a steady rain of comets into the inner solar system.

Occasionally, a star comes so close to the Sun it passes completely through the Oort cloud, violently disrupting comets in its vicinity. Some comets are tossed out of the cloud into interstellar space, while others are flung toward the Sun. A shower of comets hundreds of times the normal rate would strike the planets over a period of 2 to 3 million years, possibly causing mass extinctions of species on Earth.

The solar system's journey through giant molecular clouds at the midplane of the galaxy could restrict the Sun's output and reduce Earth's insolation, or solar input. The limited sunlight would initiate climatic changes that dramatically influence life on the planet. However, the dust clouds are not believed to be dense enough to block out the Sun significantly during each passage through the midplane, a journey that could span upward of several million years. The solar system presently resides near the midplane of the galaxy. Earth appears to be midway between major extinction events, the last two of which occurred approximately 37 million and 11 million years ago.

Since passage through the plane of the galaxy appears to have little effect, the extinction episodes might instead correlate with the approach of the solar system to its farthest extent from the galactic midplane. The extinction of the dinosaurs and large numbers of other species 65 million years ago occurred when the solar system's distance from the midplane was nearly maximum. When the solar system reaches the upper or lower regions of the galaxy, it becomes more exposed to high levels of cosmic radiation originating from supernovas (Fig. 141). The radiation would ionize Earth's upper atmosphere, producing a haze that blocks out sunlight. Furthermore, if a giant star such as Betelgeuse, 300 light-years away and 1,000 times wider than the Sun, went supernova, Earth might receive a blast of ultraviolet radiation and X rays sufficiently strong to burn off the ozone layer in the upper atmosphere. The absence of the protective ozone layer could possibly lead to the destruction of vulnerable life-forms on the planet's surface.

On a cosmic time scale, supernovas are quite frequent in this galaxy, occurring two or three times per century. They are thought to exist in two varieties. Type I supernovas lack hydrogen. They are believed to arise when a small, compact star called a white dwarf steals so much matter from a companion star that it explodes. Type II supernovas emit large amounts of hydrogen. They are thought to form when the core of a massive star collapses, generating a shock wave that triggers an explosion.

Supernovas are, for the most part, invisible because they are hidden behind dark galactic clouds. Just five supernovas are known to have been observed in this galaxy during the past 1,000 years. A supernova explosion witnessed by Chinese astronomers in A.D. 386 has been linked to a known pulsar, a rapidly spinning and highly dense neutron star. Oriental records show that in A.D. 1054, Chinese astronomers reported seeing a bright spot in the heavens that was absent before. For many weeks, the supernova could be seen even in the daytime. What they observed is now known as the Crab Nebula (Fig. 142), which is the leftover debris from the supernova.

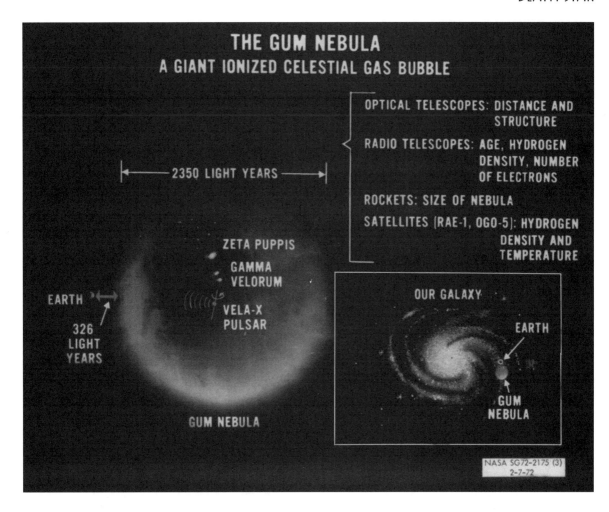

THE GUM NEBULA
A GIANT IONIZED CELESTIAL GAS BUBBLE

OPTICAL TELESCOPES: DISTANCE AND
STRUCTURE

RADIO TELESCOPES: AGE, HYDROGEN
DENSITY, NUMBER
OF ELECTRONS

ROCKETS: SIZE OF NEBULA

SATELLITES (RAE-1, OGO-5): HYDROGEN
DENSITY AND
TEMPERATURE

2350 LIGHT YEARS

ZETA PUPPIS
GAMMA
VELORUM
VELA-X
PULSAR

EARTH
326
LIGHT
YEARS

OUR GALAXY

EARTH

GUM
NEBULA

GUM NEBULA

NASA SG72-2175 (3)
2-7-72

Cassiopeia A is the remnant of a supernova some 9,000 light-years away, whose fireball was first observed in the year 1680. The fireball is presently about 10 light-years across and contains dense blobs of gases that dot the surface with blemishes as massive as 300 Earths. Stellar fragments that lagged behind punctured the expanding shell of hot gases until the shell slowed, enabling the fragments to catch up and burst through it. This effect gives the supernova its mottled appearance and might explain why some supernovas remain so bright for so long.

The first supernova observed from Earth in nearly 400 years was discovered on February 23, 1987. It was apparently once a blue giant that exploded in the Large Magellanic Cloud lying just outside this galaxy some 170,000 light-years away. Therefore, the giant star that produced the supernova would have exploded about 170,000 years ago, and its light was just reaching

Figure 141 The Gum Nebula was probably produced when the burst of radiation from a supernova heated and ionized gas in interstellar space.

(Photo courtesy NASA)

Earth. The star was of the fifth magnitude, bright enough to be seen with the naked eye for a short time.

When a giant star goes supernova, the massive nuclear explosion flings the outer layers off into space at fantastic speeds, while the core compresses into an extremely dense neutron star. The supernova releases into the galaxy large amounts of helium and radiation, consisting of deadly cosmic rays, the most energetic radiation known. The flux of cosmic rays bombarding Earth (Fig. 143) consists of atomic nuclei, protons, electrons, gamma radiation, and X rays.

Figure 142 *The Crab Nebula from Kitt Peak National Observatory.*

(Photo courtesy NOAO)

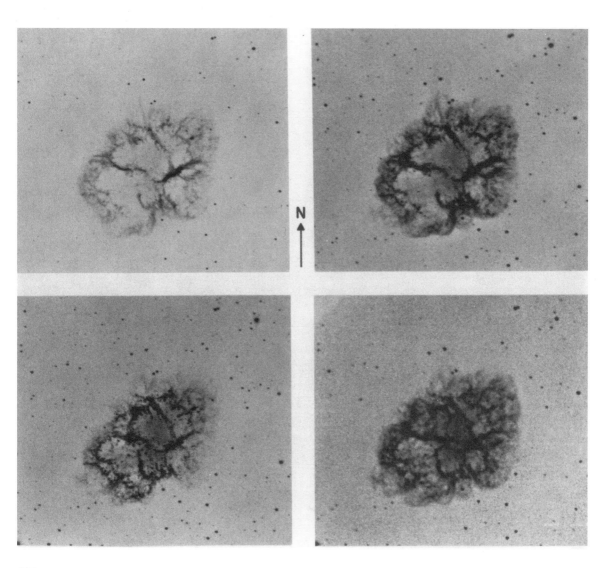

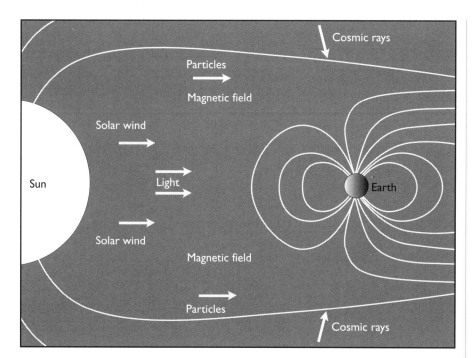

Most of the cosmic radiation presently striking the planet apparently originated from the explosion of the gigantic star Vela around 10,000 years ago and about 50 light-years from Earth. Simultaneously, large mammals, including woolly mammoths (Fig. 144), saber-toothed tigers, giant sloths, dire wolves, and mastodons, began to go extinct. Vela is believed to have emitted a burst of gamma radiation and X rays strong enough to destroy upward of 80 percent of Earth's ozone layer, allowing the Sun's harmful ultraviolet radiation to penetrate the atmosphere and kill off the vegetation that the large mammals depended on for survival.

DOOMSDAY COMETS

The French scientist Pierre de Maupertuis proposed as early as 1750 that comets had occasionally struck Earth, substantially altering the atmosphere and ocean to cause extinctions. Comets comprising ice and rocky materials range up to several tens of miles in diameter and travel in highly elliptical orbits that take them within the inner solar system. If Earth is carried into the path of one or more of these icy visitors from outer space, the collision could be devastating to life on this planet and cause almost instantaneous extinctions.

The impact on Earth of a comet-sized body creating a 100-mile-wide crater would kill everything within sight of the fireball. The pall of debris flung high into the atmosphere would place the world into a deep-freeze for several months, killing off temperature-sensitive species not protected from the cold. The extreme cold during the period of darkness would have a similar effect as if all species had been transported to present-day Antarctica.

A massive comet shower, involving thousands of impactors over Earth, might help explain the disappearance of species throughout geologic history. Comet showers are thought to recur every 100 million years or so on average and last up to 1 million years duration. Every time Earth in its orbit around the Sun is carried into the comet debris field, the possibility of multiple impacts greatly increases. The debris from the broken-up comet would then pummel Earth year after year.

The impact of an asteroid or comet that supposedly ended the Cretaceous period, abbreviated K for the German word *Kreide,* appears to have been a unique event in the geologic past. The iridium enrichment in the sediments at the Cretaceous-Tertiary (K-T) boundary was 160 times that normally found in Earth's crust. The amount of iridium spread evenly around Earth, estimated at approximately 200,000 tons, yielded the size of the impactor to be roughly 6 miles in diameter. Comets, on the other hand, contain lower concentrations of iridium than most asteroids but can strike the planet at higher speeds, thereby producing a larger impact for the same amount of iridium.

Unlike asteroids, comets at times arrive in bunches, often breaking up far out in space. For example, over the last 150 years, some two dozen comets

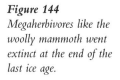

Figure 144
Megaherbivores like the woolly mammoth went extinct at the end of the last ice age.

have been observed to break up as they neared the Sun. About 20 large fragments of Comet Shoemaker-Levy 9 slammed into the far side of Jupiter beginning on July 16, 1994, sending massive plumes far above the planet's atmosphere. A similar impact on Earth would ensure extinctions of massive proportions, radically altering the course of evolution for millions of years.

As they streak toward the surface, comets shock heat the atmosphere by the expanding fireball, combining nitrogen, oxygen, and water vapor to form a strong nitric acid rain. Essentially pure nitric acid would pour over some 10 percent of the global surface in the first few months. The acids could upset the ecological balance by altering the acid-base composition of the environment. The deluge of acid would cause a massive die-off of species because most organisms cannot tolerate high acidity levels in their surroundings.

A massive bombardment of meteors or comets would strip away the ozone layer in the upper atmosphere, leaving species on the surface vulnerable to the Sun's deadly ultraviolet rays. The increased radiation would kill land plants and animals along with primary producers in the surface waters of the ocean. Indeed, many terrestrial plant species rapidly died out at the end of the Cretaceous. Plankton, which are small floating plants and animals in the sea (Fig. 145), experienced the highest rates of extinction of any group of marine organisms, with 90 percent disappearing within half a million years.

KILLER ASTEROIDS

Perhaps as many as 10 or more major asteroids have struck Earth within the last 600 million years, many of which correlate with mass extinctions of species. Meteorite impacts throughout Earth's history have produced about 150 known craters scattered throughout the world (Fig. 146). Major meteorite impacts also appear to be somewhat periodic, recurring every 26 to 32 million years.

Almost instantaneous extinctions would result from the impact of a large extraterrestrial body. A massive asteroid striking Earth with a force equal to 1,000 eruptions of Mount St. Helens would send aloft some 500 billion tons of Sun-blocking sediment into the atmosphere. The impact could also excavate a crater deep enough to expose the molten rocks of the mantle, creating an enormous volcanic eruption. Along with huge amounts of dust generated by the impact itself, large quantities of volcanic ash injected into the atmosphere could choke off the Sun.

Billions of red-hot bits of impact melt speeding through the atmosphere from suborbital trajectories would heat the surface and ignite global forest fires. Moreover, a drop in temperatures during the so-called "impact winter" would kill much of the world's forests and leave them in a desiccated and flammable state highly susceptible to ignition by lightning. The wildfires would consume a

Figure 145
Foraminiferans of the North Pacific Ocean.

(Photo by P. B. Smith, courtesy USGS)

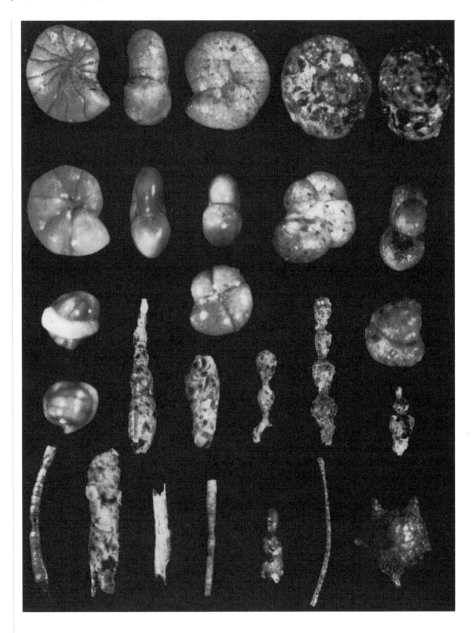

large portion of the land surface, decimating wildlife habitats and causing extinctions of massive proportions. The added atmospheric pollution would further cool the planet and halt photosynthesis, killing off species in tragic numbers.

The world would suffer from a year or more of darkness under a thick smog of soot from massive forest fires and other poisonous air pollutants such as nitrogen oxide, which is highly toxic to growing plants and air-

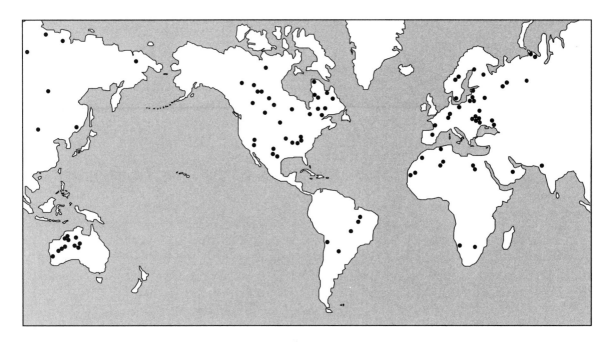

breathing animals. Global rains would contain strong acids that would percolate into the soil and rocks and would contaminate the runoff with deadly trace metals. The ocean poisoned by stream runoff and impact fallout would become a watery graveyard for all but the hardiest creatures living on the deep ocean floor.

Yet some species would survive the onslaught. Plants existing as seeds and roots would be relatively unscathed. The high acidity levels in the ocean would dissolve the calcium carbonate shells of marine organism, causing them to die out. In contrast, species with silica shells such as diatoms (Fig. 147) would survive intact. Land animals living in burrows would be well protected and would generally survive the impact, as would creatures living in lakes buffered against the acid.

The K–T boundary rocks (Fig. 148) contain a thin layer of fallout material composed of mud that settled out possibly over a period of a year or more. The mud layer is present in many parts of the world, with the greatest concentration in central North America. Within this sediment layer are shock-impact sediments, spherules, organic carbon possibly from forest fires, a mineral called stishovite found only at impact sites, meteoritic amino acids, and an unusually high content of iridium, a rare isotope of platinum in relative abundance on asteroids and comets. The iridium is not uniformly distributed throughout the world because microbes can either enhance or erase the iridium concentration in the rock. Bacteria cause the iridium to enter solution, suggesting that microbes could have erased part of the original iridium layer or spread it to deeper rock strata.

A controversy remains whether the iridium originated from an asteroid or comet impact or from massive volcanic eruptions, which are also major sources of iridium. However, volcanoes do not produce the type of shock-impact features on sediment grains such as those found at known impact sites. The spherules at the K–T boundary appear to have been created by impact melt and not by volcanism. Stishovite is a dense form of quartz that breaks down at about 300 degrees Celsius, far below temperatures generated by volcanoes. Therefore, the mineral must have had an impact origin.

Figure 147 *Diatoms from the Choptank Formation, Calvert County, Maryland.*

(Photo by G. W. Andrews, courtesy USGS)

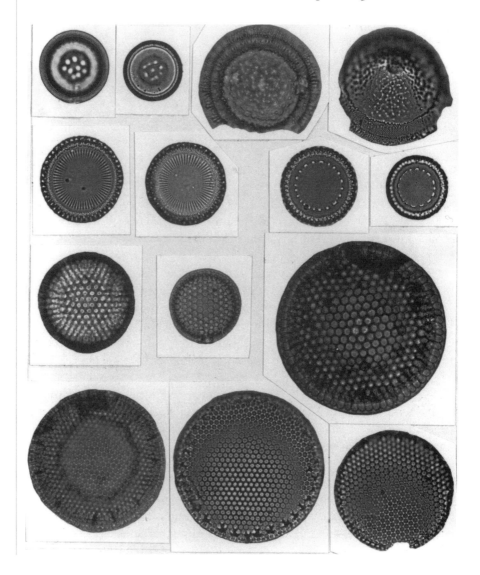

Meteorites also contain about 55 different amino acids, only 20 of which are used by living organisms. Both left-handed and right-handed molecules are created in space, but only the latter forms are made biologically. A mystery remains as to how the meteoritic amino acids managed to escape destruction from heat generated by the impact or from ultraviolet rays after they settled on the surface along with the rest of the impact fallout. The absence of amino acids in the fallout clay suggests they originally sat at the K-T boundary layer and subsequently migrated to the surrounding carbonate rocks.

The geologic record holds clues to other meteorite impacts associated with iridium anomalies that coincide with extinction episodes. Only one other iridium layer in the geologic record is as definite as the K-T boundary layer, however. It resides in the rocks of the late Eocene epoch some 37 million years ago. Another weaker iridium layer exists in rocks formed at the end of the Permian, which witnessed the greatest extinction of species on Earth (Fig. 149). Yet no other iridium concentration is nearly as prominent as the one at the end of the Cretaceous, which is as much as 1,000 times background levels, suggesting the K-T event was unique in the history of life on Earth.

Figure 148 *Geologists point out the Cretaceous-Tertiary boundary at Browie Butte outcrop, Garfield County, Montana.*

(Photo by B. F. Bohor, courtesy USGS)

Figure 149 *The number of families through time. Note the large dip during the Permian extinction.*

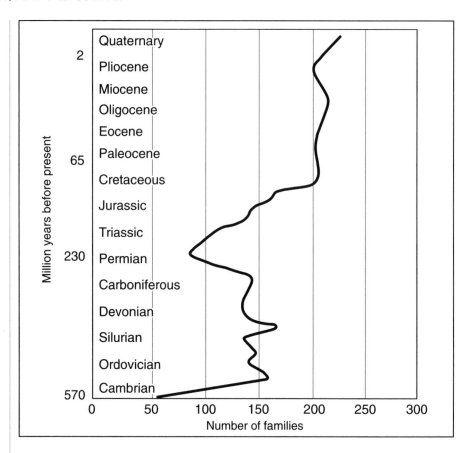

THE DINOSAUR DEATHBLOW

At the end of the Triassic period about 210 million years ago, an asteroid landing in present-day Quebec, Canada, blasted out a crater up to 60 miles wide known today as the Manicouagan impact structure. Shocked quartz grains in sediments of the Triassic-Jurassic (Tr-J) boundary are similar to those found in sediments from the K-T boundary. Three closely spaced layers of shocked quartz in the Italian Alps suggest that a shower of comets struck Earth over a period of a few thousand years.

The Tr-J boundary marks one of the most severe die-offs in Earth's history. Nearly half the ancient reptile families went extinct, which might have helped the dinosaurs rise to dominance. The dinosaurs were no doubt the most successful land creatures that ever lived and ruled the world for 140 million years. Their extensive range, wherein they occupied a wide variety of habitats and dominated all other forms of land-dwelling animals, exemplifies the success of the dinosaurs.

Many hypotheses have been presented to explain the demise of the dinosaurs, including massive volcanic eruptions, widespread diseases, and even the evolution of flowering plants, which drew down the oxygen content in the atmosphere. The fact that the dinosaurs were not the only ones to go extinct and that 70 percent of all known species and 90 percent of all biomass died out at the end of the Cretaceous indicates something in the environment went terribly wrong.

One popular theory for the extinction of the dinosaurs and many other species living during the time contends that a meteorite 6 miles wide hit Earth. The huge meteorite excavated a deep crater 100 miles or more across located off the Yucatán peninsula near the small town of Chicxulub in the Gulf of Mexico. The meteorite slammed into a relatively rare rock type that made up a thick carbonate platform comprised of sulfur-rich limestone. The crash filled the atmosphere with an aerosol of sunlight-blocking sulfuric acid, which dropped onto land, defoliating plants, and fell into the oceans, turning the surface waters into a toxic soup. Carbon dioxide liberated during the crash produced a runaway greenhouse effect that turned the planet into a hot oven.

The giant asteroid or comet apparently struck Earth from the southeast at a velocity of about 10 miles per second and at a shallow angle of around 30 degrees or so. A shallow approach would also have created a far larger fireball than a more vertical one and would be particularly devastating to the interior of North America as indicated by thick deposits of tektites. The explosion was so powerful that impact debris might have been flung as far away as the North Pacific.

The enormous energy liberated by the impact would have created many environmental disasters, including mile-high tsunamis, massive storms, cold and darkness followed by powerful greenhouse warming, strong acid rains, and global forest fires. When conditions returned to normal within a few years, more than half the floras and faunas, including the dinosaurs (Fig. 150), were missing. Earth had not experienced such an event since complex life first appeared 1 billion years ago. Geologic history had therefore taken a new and unexpected turn.

Instead of a single large bullet, one or more asteroids or comets might have struck Earth. However, simultaneous impacts with asteroids are rare, whereas comets often strike in groups. The impact would have the equivalent explosive force of 100 trillion tons of TNT, more powerful than the detonation of 1,000 times all the nuclear arsenals in the world. Such a bombardment would send a half trillion tons of debris flying into the atmosphere and spark planet-wide blazes that would burn perhaps one-quarter of all vegetation on

Figure 150 *Triceratops were among the last dinosaurs to become extinct at the end of the Cretaceous.*

the continents. This would plunge the entire Earth into an ecologic calamity. The combination of the dust and smoke would reduce surface temperatures several degrees for many months.

After the impact and the ensuing cooling period, greenhouse warming from excess water vapor and carbon dioxide in the atmosphere would roast the planet. Many plants and animals, including the dinosaurs, might have survived the severe cold only to lose their lives to a subsequent period of extreme heat. The fossil record supports this contention and indicates that ocean temperatures actually rose by 5 to 10 degrees Celsius for tens of thousands of years beyond the end of the Cretaceous.

More than 90 percent of microscopic marine plants, called calcareous nannoplankton went extinct along with most marine life in the upper portions of the ocean for almost half a million years. The plankton might have been killed off by a lack of sunlight needed for photosynthesis or by acid rain following a massive meteorite impact. The acid in the ocean might also have been sufficiently strong to dissolve the plankton's calcareous shells, causing extinctions of unprecedented levels.

The death of the calcareous nannoplankton could have raised global temperatures sufficiently to kill off the dinosaurs and other species. The plants produce a sulfur compound that aids in cloud formation. The clouds, in turn, reflect sunlight and prevent solar radiation from reaching the surface. Instead of a climate change causing an extinction, perhaps the extinction of the nannoplankton dramatically affected the global climate.

The moist greenhouse effect (Fig. 151), causing temperatures to rise by increased carbon dioxide levels, generated by the impact might have transformed savannas into rain forests, drastically upsetting the environment. A comparison between the shapes of modern leaves and leaf fossils found in sediments before and after the time of the proposed impact has uncovered an abrupt transition from small, rounded leaves to large, pointed ones, suggesting precipitation and temperatures soared at the end of the Cretaceous. Forests containing entirely different floras replaced forests of broad-leaved trees and shrubs living before the K–T boundary.

Warmer ocean temperatures could spawn super hurricanes that reached altitudes of 30 miles, high up into the stratosphere, the upper level of the atmosphere, (Fig. 152) and more than twice the altitude of normal hurricanes. Evidence suggests that a large extraterrestrial body landed on carbonaceous rocks in the Gulf of Mexico at the end of the Cretaceous period and injected massive quantities of greenhouse-warming carbon dioxide into the atmosphere. The impact would have warmed the water 50 degrees Celsius, almost double current temperatures in the tropics, adding substantially more power to hurricanes. Because of their great height, the tropical storms could have transported water vapor, ice particles, and dust into the stratosphere, where they would block out sunlight and destroy the life-protecting ozone layer.

The massive meteorite bombardment itself could also have stripped away the upper stratospheric ozone layer, thus bathing the planet in the Sun's

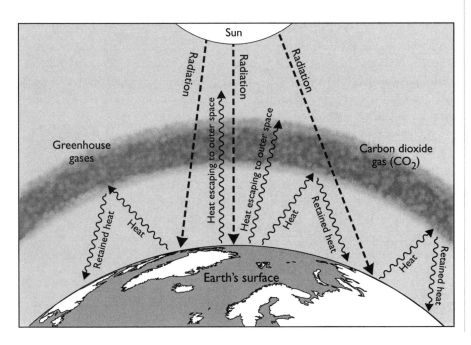

Figure 151 The greenhouse effect: solar radiation passing through the atmosphere is converted on Earth's surface into infrared radiation, which escapes upward, is absorbed by greenhouse gases, and is reradiated toward Earth.

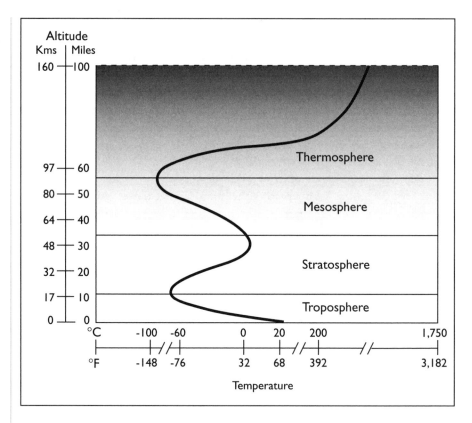

Figure 152 The layering of the atmosphere.

deadly ultraviolet rays. The radiation would kill land plants and animals along with primary producers in the surface waters of the ocean. On land, the onslaught of ultraviolet radiation, occurring during only the daytime, would have spared the nocturnal mammals, while the naked dinosaurs were left fully exposed out in the open.

The impact might have lofted billions of tons of debris into orbit around Earth, leaving the planet with Saturnlike rings, which would have taken about 100,000 years to form. The rings would block out the Sun and cast a deep, 700-mile-wide shadow across the surface, much like a total eclipse. The shadow would have turned warm tropical rain forests into colder, temperate regions, devastating warmth-loving faunas and floras. The rings would eventually disappear after 2 or 3 million years as the orbiting debris fell to Earth and burned up in the atmosphere. The bombardment of the atmosphere would have wreaked havoc with the climate and endangered the lives of whatever life survived the meteorite impact. Ironically, while the rings lasted, they would have made Earth the most beautiful planet in the solar system.

NEMESIS

One explanation for the sudden mass extinctions that seem to recur in the fossil record every 26 to 32 million years over the past 250 million years (Fig. 153) envisions a hypothetical companion star of the Sun named Nemesis in honor of the Greek goddess who dishes out punishment on Earth. Twin star systems are common in the Milky Way galaxy. More than half of all known stars are binary or multiple systems. As with other binary stars, Nemesis could have formed much closer to the Sun 4.6 billion years ago and slowly spiraled outward due to stellar perturbations similar to the manner in which the comets in the Oort cloud did. The Sun could also have captured the tiny star more recently, perhaps within the last 600 million years.

Nemesis is thought to be a brown dwarf. Because of its small size, it could not ignite into a full-blown star and therefore would not be easily seen from Earth. Nemesis is believed to be more than a light-year away. Because of its dim light at this distance the star proves difficult to find even with the most powerful telescopes. It apparently revolves around the Sun in a highly elliptical orbit inclined steeply to the ecliptic and spends most of its time outside the Oort cloud, a repository of several trillion comets that surrounds the solar system just inside the orbit of Nemesis.

Its orbital motion would enable Nemesis to approach the Oort cloud about every 26 million years as it passes through perihelion, the point of

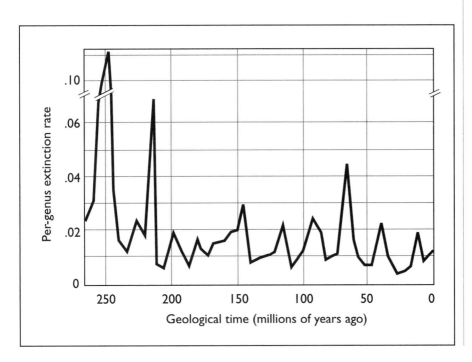

Figure 153 Species extinctions, beginning with the great Permian die-off 250 million years ago.

closest approach to the Sun. As Nemesis cuts a path deep into the comet cloud, its gravitational disturbance would distort the orbits of comets in its vicinity and send a million-year storm of comets, involving some 100 million objects, into the inner solar system. Furthermore, the number of comets could be greatly enhanced if they interacted with a wide band of comets in the Kuiper belt.

The disk of the solar system does not end abruptly at Neptune or Pluto, which compete with each other as the most-distant planet from the Sun. Instead, a wide belt of comets lies beyond Neptune and Pluto in the plane of the solar system, consisting of residual material left over from the formation of the planets. The density of matter in this outer region is too low for the formation of large planets, but many smaller bodies perhaps the size of asteroids appear to have accreted in this area.

The Kuiper belt contains at least 35,000 bodies larger than 60 miles wide, giving it a mass several hundred times larger than the asteroid belt between Mars and Jupiter. These scattered remnants of primordial material are so far from the Sun that they are quite cold. Therefore, these distant objects are likely composed of water ice and frozen gases, making them similar to the nuclei of comets. The interaction of comets on the inner edge of the Kuiper belt with Neptune's gravitational attraction flings them into the interior of the solar system.

As the swarm of comets heads toward the Sun, many inevitably rain down on Earth when its orbit intercepts the paths of these icy invaders. With 1 billion Earth-crossing comets cascading in from the inner Oort cloud, about 20 large impactors up to 1 mile or more wide and numerous smaller ones would be expected to strike the planet over an interval of 2 to 3 million years. The comets raining down onto Earth could wreak considerable damage, forcing the extinction of large numbers of species. Even the prelude to the K–T disaster, when dinosaurs were already in decline, might have been a 1-million-year rain of smaller comets punctuated by a single knockout punch.

The apparent periodicity of the mass extinctions also seems to coincide with the ages of terrestrial impact craters. Moreover, the periodicity of the ages of the H class of meteorites, which are chondrites with a high iron content, coincides with times of mass extinction. The meteorites are presumed to be pieces broken off of asteroids by the impacts of other asteroids or comets.

Large numbers of meteorite liberations crowd at or near mass extinction events. Such widescale meteorite formation could result from showers of comets moving through the asteroid belt. The comets hitting Earth could trigger climate changes that cause mass extinctions. The inevitable rain of objects onto the planet could therefore change the rules governing the evolution and extinction of species.

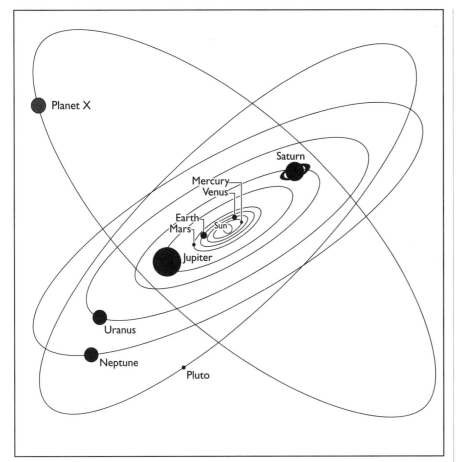

Figure 154 *The supposed location of Planet X with respect to the solar system.*

PLANET X

In the depths of space far too dim to be seen by the most powerful telescopes is thought to be a tenth planetary body dubbed Planet X (Fig. 154). The idea that an extra planet existed on the periphery of the solar system has been put forth several times over the last 100 years to explain observed deviations in the motions of the outer planets from their predicted courses. The encounters also would make the orbit of Planet X extremely unstable due to the gravitational nudges from other planetary bodies.

Planet X is believed to lie well outside the orbit of Pluto, perhaps 10 billion miles from the Sun. The elusive planet apparently revolves around the Sun in an elongated orbit steeply inclined to the ecliptic and takes perhaps 1,000 years to complete one revolution. The possibility also exists that Planet X was simply a one-time visitor to the solar system and does not circle the Sun at all.

Planet X is apparently no larger than five Earth masses because a much larger body would have been identified by now. The present model of the solar system is not sufficiently complete, however, to predict the planetary motions with the level of accuracy required to detect Planet X. The presence of Planet X might be ascertained indirectly by its gravitational influence on Uranus and Neptune, which were deflected from their paths around the Sun during the 19th century. A similar gravitational effect on the orbital motions of Uranus and Neptune led to the discovery of Pluto in 1930. Yet no deflection of Uranus and Neptune by Planet X has been detected during the 20th century, which suggests it must be in a peculiar orbit.

Gravitational perturbations by the outer planets could cause Planet X to precess at just the proper rate to bring it into the Kuiper belt, a flattened disk of comets outside the orbit of Neptune, every 26 million years or so. Planet X would have cleared a gap in a section of the belt that is closest to the Sun. The gravitation pull of Planet X would dislodge comets near the gap when it approaches the inner and outer edges of the gap. Between the time the body approaches and departs the comet belt, its gravity would disturb comets within its vicinity, sending a shower of icy objects into the inner solar system, some of which might rain down havoc on Earth over an extended period.

After discussing the periodic impact extinction of species, the next chapter examines asteroid and comet bombardment of Earth and its aftermath.

10

COSMIC COLLISIONS
ASTEROID AND COMET BOMBARDMENT

This chapter examines the effects on civilization caused by the bombardment of large bodies from outer space. The ultimate environmental threat to the human race would be the devastation resulting from the collision of asteroids or comets with this planet. Many Earth-bound asteroids and comets capable of killing 1 billion or more people are wandering around in space. Hundreds of these large, near-Earth asteroids have been discovered over the decades. Occasionally, a wayward object comes close enough to cause alarm. An impact of such a body would certainly jeopardize humanity.

Large meteorite craters located throughout the world are testimony to disasters of the past caused by major meteorite impacts. Should such an impact occur today, it would create as much damage as global nuclear war. Indeed, the environmental consequences would be similar to those of nuclear winter, making survival difficult for life on Earth and radically changing the course of human history.

NEAR-EARTH ASTEROIDS

About 200 near-Earth asteroids (NEAs) ranging up to 25 miles wide have been observed to be out of the main asteroid belt and in Earth-crossing orbits. Moreover, from several hundred to as many as 2,000 asteroids wider than half a mile are poised to cross Earth's orbit for a close encounter. So far, more than 100 objects have been found that might pose a hazard to the planet, representing only one-tenth of the estimated total. The first Earth-crossing asteroids, whose paths around the Sun intersect this planet's orbit, were discovered in 1932. Since then, hundreds more have been discovered.

Interestingly, an NEA called 719 Albert, observed several times in 1911, had not been seen since until May 1, 2000. By 1940, astronomers had also lost track of many other known asteroids. By the 1970s, they had rediscovered all but 20 of them. By 1991, only Albert remained unaccounted for. Asteroid Albert has an orbital period of 4.3 years and comes within 20 million miles of Earth. The most distant point of its orbit is about 260 million miles from this planet, a distance that might have prevented it from being rediscovered any earlier.

How these asteroids managed to obtain orbits that intersect Earth's path remains a mystery. Apparently, they run in nearly circular courses for 1 million or more years. Then, for unknown reasons, possibly due to the gravitational attraction of a passing comet or the pull of the giant planet Jupiter (Fig. 155), their orbits suddenly stretch and become so elliptical that some asteroids come within reach of this planet. Regardless, the likelihood of a collision with a large asteroid appears to be quite remote, given the great distances between planetary bodies.

An asteroid 1 or 2 miles wide aimed at Earth would slam into the atmosphere at 100 times the speed of a high-powered bullet. Seconds later, it would strike the planet with an explosive force of more than 100,000 megatons of TNT. The shock wave from the impact would level everything within several tens of miles and launch a plume of vaporized and pulverized rock high up into the atmosphere. Debris would then fall back to Earth, peppering the ground with rocks ranging in size up to boulders.

The near-Earth asteroids are called Apollos, Amors, and Atens. Apollos cross Earth's orbit, and some have come within the distance of the Moon. Amors cross the orbit of Mars and approach Earth, perhaps crossing its orbit in the course a few hundred to a few thousand years. Atens spend most of their time inside Earth's orbit and might intersect the planet's path at their aphelion, or farthest distance from the Sun.

These bodies are not confined to the asteroid belt as are the great majority of known asteroids. Instead, they approach or even cross the orbit of Earth. The asteroids might have begun their lives outside the solar system and appear to be former comets. After repeated encounters with the Sun, they have exhausted their volatile materials and lost their ability to produce a coma or tail. Throughout eons, their coating of ices and gases has been eroded away by the Sun, exposing what appears to be large chunks of rock.

Figure 155 *Jupiter and its four planet-sized moons, from* Voyager 1 *in March 1979.*

(Photo courtesy NASA)

A number of Apollo asteroids have been identified out of a possible total of 1,000. Most are generally small and discovered only when swinging close by Earth (Table 11). Usually, within a few million years, they either collide with one of the inner planets or are flung out in wide orbits after a near miss. Inevitable collisions with Earth and the other inner planets steadily depletes their numbers, requiring an ongoing source of new Apollo-type asteroids either from the asteroid belt or from the contribution of burned-out comets.

The asteroid Phaethon, named for the mythical son of Helios the Greek Sun god, approaches the Sun within the shortest distance of any known asteroid. It has a perihelion distance of only 13 million miles, some 20 million miles closer to the Sun than Mercury. Phaethon also has the smallest orbit of any known comet. It is rather large compared with the esti-mated diameters of most observed comets, with a diameter of 3 to 4 miles. This is comparable to the size of most near-Earth asteroids. Its dark color and rapid rotation also seem to exclude it from being a former comet. Most asteroids rotate with periods of between two and 60 hours, with an average near eight hours. Furthermore, a rapidly rotating comet would soon spin off its loose outer layers.

Phaethon presently crosses the plane of the solar system just outside Earth's orbit (Fig. 156), therefore no collision is imminent. However gravi-tational influences are slowly moving this crossing point outward. Therefore, in 250 years, the two orbits will intersect, possibly resulting in alarmingly close encounters with Earth. Perhaps the day this inevitable event occurs, Phaethon might briefly reach naked-eye visibility. Fortunately, only one such near collision can occur, because the asteroid's proximity to Earth will radically alter Phaethon's orbit and prohibit any further close encounters with the wayward asteroid.

To track these near-Earth asteroids, the Spacewatch Telescope was estab-lished at Kitt Peak National Observatory near Tucson, Arizona, in the early 1970s. It employs a 36-inch telescope originally built in 1919 fitted with highly sensitive solid-state light detectors called charged-coupled devices connected to

TABLE 11   CLOSEST CALLS WITH EARTH

Body	Distance in angstrom units (Earth-Moon)	Date
1989 FC	0.0046 (1.8)	Mar. 22, 1989
Hermes	0.005 (1.9)	Oct. 30, 1937
Hathor	0.008 (3.1)	Oct. 21, 1976
1988 TA	0.009 (3.5)	Sep. 29, 1988
Comet 1491 II	0.009 (3.5)	Feb. 20, 1491
Lexell	0.015 (5.8)	Jul. 1, 1770
Adonis	0.015 (5.8)	Feb. 7, 1936
1982 DB	0.028 (10.8)	Jan. 23, 1982
1986 JK	0.028 (10.8)	May 28, 1986
Araki-Alcock	0.031 (12.1)	May 11, 1983
Dionysius	0.031 (12.1)	Jun. 19, 1984
Orpheus	0.032 (12.4)	Apr. 13, 1982
Aristaeus	0.032 (12.4)	Apr. 1, 1977
Halley	0.033 (12.8)	Apr. 10, 1837

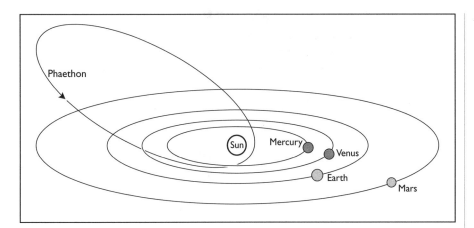

Figure 156 *The orbit of Phaethon with respect to the inner solar system.*

a powerful computer. When the telescope scans across the sky, any movement of an object compared to the background stars is registered and the orbit calculated. The National Aeronautics and Space Administration (NASA) is attempting to catalog 90 percent of all asteroids a half mile across or larger that have orbits approaching Earth. This program is called Near-Earth Asteroid Tracking (NEAT). Another effort, named the Lincoln Near-Earth Asteroid Research (LINEAR) program, can scour huge swaths of the heavens for faint, moving objects, any one of which could cause considerable damage if it struck Earth.

The closest known approach of a sizable asteroid tracked by Spacewatch occurred on December 5, 1991. It observed a 30-foot-wide object named 1991 BA that came within 100,000 miles of Earth, less than half the distance to the Moon. Three years later, a house-sized object whizzed past Earth within 65,000 miles. Many Earth-approaching asteroids, called the Arjunas, named in honor of the Indian prince in an epic Hindu poem, are quite small, with diameters of less than 150 feet. They appear to ride in circular orbits around the Sun near the path of this planet.

The unexpectedly large number of these small asteroids argues for the existence of another asteroid belt near Earth that might represent material gouged out of the Moon when giant asteroids slammed into it eons ago. The asteroidal fragments could have been nudged out of the main asteroid belt under the influence of Jupiter's gravity after repeated collisions with other asteroids. They could also be fragments of comets that passed close to Earth.

A more aggressive global detection network called Spaceguard, would consist of six 100-inch telescopes based around the world. They would be equipped with extremely sensitive, charged-coupled devices that could detect faint objects the size of asteroids far off in space. The technology would provide up to 20 years warning of Earth-approaching asteroids larger than a half mile across, which would be sufficient time to take evasive action. Lead times of 50 to 100 years would be more desirable since the approaching object would then need only a slight nudge from a rocket, for instance, to knock it off its Earth-bound course.

Unfortunately, a false warning could cause as much havoc as the asteroid itself. Asteroid 1997 XF-11 was reported to pass some 30,000 miles from Earth in the year 2028—a hair's breadth astronomically speaking. However, a further check of photographs of the object taken in 1999 determined that the asteroid would miss the planet by twice the distance to the Moon. A similar situation occurred with asteroid 2002 NT 7 in July 2002. The fear is that people might not take such close calls seriously, possibly to their detriment if indeed a large meteorite were destined to crash into Earth.

CLOSE CALLS

One of the closest encounters with a major asteroid occurred on October 30, 1937, when Hermes shot past Earth at 22,000 miles per hour. The mile-wide asteroid missed hitting the planet by a mere half million miles, or about twice the distance to the Moon—extremely close in astronomical terms. As with many asteroids, Hermes swings around the Sun in an eccentric orbit that crosses Earth's orbital path, which increases the possibility of a chance collision.

Had Hermes collided with Earth, it would have released the energy equivalent of 100,000 megatons of TNT, or 2,000 times larger than the biggest hydrogen bomb ever built. Indeed, nuclear war has many similarities to the impact of a large asteroid. The impact would send aloft huge amounts of dust and soot into the atmosphere. The debris would clog the skies for several months, plunging the planet into a deep freeze. All inhabitants would face severe hardships as temperatures plummeted and Arctic conditions prevailed over large parts of the globe.

Another close flyby of an asteroid occurred on March 22, 1989, when asteroid 1989 FC came within 430,000 miles of Earth (Fig. 157). The asteroid was about a half mile wide. If it had struck Earth, a fluke of orbital geometry might have softened the blow somewhat. The asteroid orbits the Sun in the same direction as Earth, completing one revolution in about a year. Therefore, its approach was rather slow compared with the relative motions of other celestial objects. However, because of Earth's large size, the planet's strong gravitational pull would have greatly accelerated the asteroid during its final approach.

If a collision had occurred, the asteroid would have produced a crater 5 to 10 miles wide, large enough to wipe out a major city. Had the asteroid been 10 times larger, about the size of the dinosaur killer, the impact would have caused the planet to vibrate like a giant bell. The collision could trigger strong earthquakes, powerful volcanic eruptions, great tsunamis (if landing in the sea),

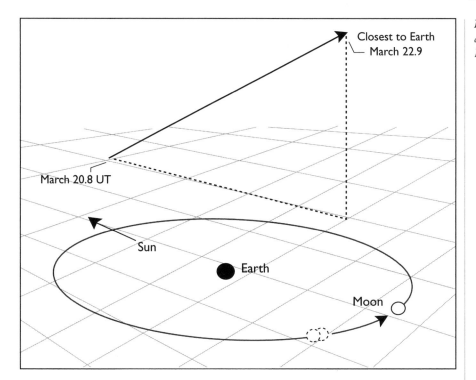

Figure 157 *The closest approach of asteroid 1989 FC to Earth.*

and global forest fires. Dust, smoke, and water vapor would contaminate the atmosphere up to a year or more.

Astronomers did not detect asteroid 1989 FC until it had already passed by Earth. Only then did they notice a dramatic slowdown in the asteroid's motion against background stars. To their amazement, the asteroid was rushing on a straight-line course away from the planet on what must have been a near-grazing trajectory. The astronomers failed to notice the approach of the asteroid because it came from a Sunward direction. Moreover, the Moon was nearly full, further hampering observations.

On December 8, 1992, a large asteroid called Toutatis, measuring 1.6 by 2.5 miles, flew within 2.2 million miles of Earth. The two irregularly shaped, cratered chunks comprise a contact-binary asteroid held together by gravity. Toutatis's orbit brings it near Earth every four years. These double bodies appear to be quite common among the Earth-approaching asteroid population.

On May 19, 1996, the largest known asteroid to come within the distance of the Moon, named 1996 JA1, with an estimated diameter of 1,000 to 1,500 feet, whizzed past Earth at a distance of about 280,000 miles. Only six recorded asteroids, each measuring less than 300 feet wide, have come closer. One was 2002 MN, which was about the size of a football field and came within a mere

75,000 miles of Earth on June 14, 2002. Instead of approaching from the plane of the solar system where most asteroids are thought to reside, 1996 JA1 orbited 35 degrees to the ecliptic, a strange place for an asteroid.

Comets have also been known to fly uncomfortably close to Earth. The closest comet to approach the planet was Lexell, which came within six times the distance to the Moon on July 1, 1770. The April 10, 837, encounter with Comet Halley was close enough for Earth's gravity to disturb the comet's orbit. On May 1, 1996, Comet Hyakutake with its long tail stretching one-third the distance across the sky achieved its closest approach to the Sun for the first time since the end of the last ice age some 10,000 years ago.

Nearly all close encounters with asteroids took astronomers completely by surprise, and not a single one was anticipated. To avoid the danger of an asteroid collision, the threatening body would first have to be tracked by telescopes and radars, and its course plotted accurately so its orbit could be determined precisely. If an asteroid were detected on a collision course with Earth, astronomers could provide timely warnings for the evacuation of the affected areas.

ASTEROID IMPACTS

The first recorded explosion of an asteroid or comet nucleus, called a bolide, occurred on June 30, 1908, in the Tunguska forest of northern Siberia (Fig. 158). As the bolide ripped through the atmosphere, the air exerted tremen-

Figure 158 *The location of the Tunguska event in northern Siberia.*

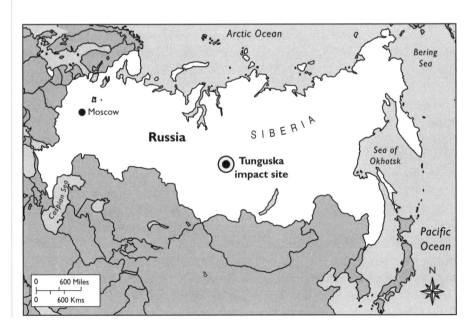

216

dous pressure on its leading side, causing the rock to deform and spread outward. The air in front of the bolide heated to tens of thousands of degrees Celsius, causing it to burn up. With practically no pressure applied to the back side, the huge differences in forces acting on various areas of the object caused it to tear into pieces. The very same forces blasted the fragments themselves apart. The bolide then shattered into a cloud of debris, as though it had been dynamited in midair.

A vast fireball raced from east to west across the skies at roughly a 45-degree angle and exploded about 4 miles above the ground with a force of 1,000 Hiroshima atomic bombs. The tremendous explosion incinerated herds of reindeer and toppled and charred trees within a 20-mile radius. Tree trunks were splayed outward from the center of the blast like the spokes of a bicycle wheel. Embedded in the trees were tiny particles having an unmistakable extraterrestrial origin. The explosion registered on scientific instruments scattered around the globe, and the sound was heard as far away as England.

Barometric disturbances were recorded over the entire world as the shock wave circled Earth twice. The dust generated by the explosion produced unusual sunsets and bright skyglows observed at night over Europe within a few days of the event. The faint red glow, produced when dust was lofted into the upper atmosphere high enough to reflect sunlight long after sunset, was bright enough so that one could read by it.

The impactor was relatively small, estimated at between 100 and 300 feet wide. This explains why no astronomical sightings were made prior to the explosion. The estimated force of the blast was as powerful as a 20-megaton hydrogen bomb. The fireball produced surface temperatures hot enough to scorch trees and other vegetation. The detonation left no impact crater or meteoritic debris, suggesting a comet or stony asteroid airburst in the lower troposphere at a speed of 40 miles per second.

A concentration of ablation products, including iron oxide and glassy spherules of fused rock was found in a tongue stretching 150 miles northwest of the impact site. The evidence indicates that the object probably arrived from a southeast direction. Furthermore, iridium was located in a peat layer about 1.5 feet below the surface at the Tunguska site. A similar concentration of iridium was discovered in samples of Antarctic ice from the same period.

The explosion of a single object over Siberia apparently could not have distributed a global veneer of debris so evenly, however, suggesting that other similar events might have occurred simultaneously in other parts of the world. The intruder's reported angle of entry indicates it could have come from a hail of cometary debris known as the Taurid shower, through which Earth passes every June and November.

The Tunguska event was the most violent impact of the 20th century. Had the impactor exploded over a sizable city, the entire town and its suburbs would be laid waste, similar to the atomic bombing of Hiroshima, Japan, at the end of World War II (Fig. 159). Earth can expect to incur a similar asteroid or comet explosion about once every 100 years. Therefore, another such impact would be expected in the not-too-distant future. Despite decades of searching, astronomers have not observed an object as small as the Tunguska impactor. Yet thousands of these objects are thought to exist within easy reach of the planet.

Of the eight small objects known to pass by Earth, half would cause Tunguska like explosions upon hitting the atmosphere. For instance, a near miss occurred on August 10, 1972, when an asteroid estimated at 260 feet wide sped through the upper atmosphere over North America, blazing across the sky in a daylight fireball before reentering space. The asteroid was visible for nearly two minutes as it traveled 900 miles between Salt Lake City, Utah, and Calgary, Alberta, Canada, coming within 35 miles of the ground before bouncing off the atmosphere.

The first explosion of an asteroid observed in recent times was reported on April 9, 1984, by the pilot of a Japanese cargo plane inbound for Alaska

Figure 159 *The destruction of Hiroshima, Japan, following the atomic bombing on August 6, 1945.*

(Photo courtesy Defense Nuclear Agency)

over the Pacific Ocean about 400 miles east of Tokyo, Japan. The round ball cloud rapidly ballooned outward similar to that of a nuclear detonation (Fig. 160), only no fireball or lightning usually associated with such explosions was visible. Yet several pilots did report seeing a slight luminosity of the spherical cloud as it rose from the cloud deck.

The cargo plane encountered only minor turbulence and experienced no communication breakup or any other problems with the aircraft's instruments often caused by a nuclear blast. The mushroom cloud expanded to 200 miles in diameter, rising from 14,000 to 60,000 feet in a mere two minutes, traveling about 260 miles per hour. At the beginning, the cloud appeared opaque. As it continued to grow, however, it became more transparent. The entire episode lasted a little less than an hour.

Dust samples collected at the scene by a military aircraft dispatched from Japan contained no radioactivity. Nor was radioactivity found on the cargo

Figure 160 A spherical cloud from the detonation of a hydrogen bomb in the Marshall Islands on November 1, 1952.

(Photo courtesy U.S. Navy)

219

plane or three other aircraft that had flown near the mushroom cloud. Furthermore, undersea instruments located near Wake Island close to the mushroom cloud site failed to detect an undersea nuclear explosion.

Since active submarine volcanoes are known to exist in the area (Fig. 161), an increase in volcanic activity on April 8 and 9 implied that an undersea volcano might have violently erupted and produced the suspicious cloud. One volcano in question, named the Kaitoku Seamount, is located 80 miles north of Iwo Jima and 900 miles southwest of the mystery cloud. However, the winds at that time were blowing in the wrong direction for the volcanic activity to have been the source of the cloud.

An intriguing alternative explanation suggests an exploding asteroid or comet 80 feet in diameter, which released the equivalent energy of a 1-megaton hydrogen bomb, produced the cloud. When the meteor encountered the cloud

Figure 161 *Submarine eruption of Myojin-sho Volcano, Izu Islands, Japan.*

(Photo courtesy USGS)

deck, it instantaneously shattered. The kinetic energy was converted into heat that evaporated cloud particles, with the hot gasses forming a rapidly rising plume. Because of the meteor's high speed, the shattering occurred over a distance of several miles, thus creating a large plume accompanied by little of the barometric disturbances normally linked with a nuclear explosion. A celestial object of this size is expected to enter the atmosphere once every 30 years.

Satellites have also uncovered unusual atmospheric disturbances. A mysterious flash over the Indian Ocean near Prince Edward Island off South Africa was detected by the Navy's surveillance satellite *Vela* on September 22, 1979. The event was thought to be atomic bomb testing by South Africa, which had long been suspected of building nuclear weapons. Another bright flash near South Africa in December 1980 prompted further investigation, although no clear physical proof was found to implicate a nuclear detonation. A panel of independent scientists reviewed all the collected data and surmised that the blast resulted from a natural event such as an exploding meteor or comet.

ASTEROID DEFENSE

Perhaps the precedent for establishing an asteroid defense system was set during the Cold War in the late 1960s when the United States initiated a ballistic missile defense system called Spartan (Fig. 162), developed to protect American cities from Russian ICBMs. The Icarus Project, named for the asteroid Icarus nearly 1 mile in diameter, was initiated by the Massachusetts Institute of Technology in 1967. Scientists conducted a study on methods to protect Earth against a collision with a major asteroid. One concept was to launch nuclear missiles and detonate them alongside the asteroid to deflect it from its Earth-bound course. This strategy was thought to have a 90 percent chance of succeeding to steer an asteroid away from Earth, provided the encounter took place far out in space.

One contention holds that even asteroids only 30 feet across should be deflected when approaching Earth closer than the distance of the Moon. The explosion of such a body on impact would be as powerful as an atomic bomb detonation that could level a large city. An asteroid 80 feet in diameter, with an impact energy equivalent to a 1-megaton hydrogen bomb, is thought to enter Earth's atmosphere about once every 30 years. However, since Earth is 70 percent ocean, most simply fall into the sea.

Of the 15,000 numbered asteroids, an estimated 1,000 large, planet-crossing asteroids are poised to brush past Earth. As many as five new ones are discovered yearly. Current space technology can turn away a menacing asteroid by setting off nuclear explosives around it. However, the detonations must not break up the asteroid. If they did, what otherwise would have been a sin-

gle bullet striking the planet might turn into buckshot, greatly aggravating the impact effects and causing considerably more damage.

Although flying rubble piles, which make up a large portion of asteroids, seem less threatening than huge solid rocks, they would wreak as much havoc if they hit Earth and might be much harder to fend off. Researchers looking for applications for technology developed for the Strategic Defense Initiative, otherwise known as "Star Wars," have proposed shattering and dispersing oncoming asteroids with nuclear missiles and other space technology (Fig. 163). However, breaking up piles of rubble is much harder than thwarting large rocks.

An alternative method of deflecting rogue asteroids on collision course with Earth is to turn away threatening asteroids or comets by the nearby detonation of a neutron device. This is essentially a hydrogen bomb without the outer uranium shell that produces an enormous explosion, thereby producing a powerful force that does not shatter the asteroid. Simply blowing up the approaching menace would only result in fragmentation, which would yield a deadly swarm of city-killer objects that remain on collision course.

One idea is to nudge the threatening body off its deadly path without shattering it by detonating a neutron bomb off to the side of the asteroid. The warhead would leave the object intact, while the intense radiation would heat the asteroid's surface sufficiently to vaporize it. The resulting jet of vapor would act like a rocket engine that provides enough thrust to deflect the object from its collision course. Such phenomena have been observed on comets, as jets of gas off to one side seem to steer the object off its normal trajectory.

Initially, the threatening object would have to be tracked with telescopes and radars and its course computed as accurately as possible so its orbit could be determined precisely. Wide-field photography has long been the most useful tool for tracking asteroids. However, for most asteroids smaller than 20 miles in diameter, little is known other than their brightness and orbit around the Sun. Furthermore, the advent of radar has made pinpointing solar system objects substantially easier. The radar signals bounce off each body and yield their distance as well as velocity.

The detonation of powerful nuclear weapons against an asteroid must be conducted as far out in space as possible, preferably beyond the orbit of the Moon. At the instant a nuclear weapon is detonated in space, atoms surround-

Figure 163   *An artist's concept of an antisatellite missile launched from an F-15 aircraft.*

(Photo courtesy U.S. Air Force)

ing the fireball are stripped of their electrons by high-intensity gamma rays. The electrons shoot out at tremendous speeds, producing extremely powerful electric fields that generate a strong pulse of radio waves, called an electromagnetic pulse (EMP), which can blanket an area several hundred miles wide. Any electrical conductor can overload and short-circuit, disabling electronic equipment, including telephones, radios, radars, devices, and computers.

Moreover, the farther out the interception is made, the greater the likelihood of deflecting an asteroid from its Earth-bound trajectory. This is because even a slight course change far out in space could steer the asteroid sufficiently to cause it to miss Earth by a wide margin. Otherwise, if the asteroid were allowed to approach too close, nuclear deflection might be ineffective, and the resulting impact could be catastrophic.

IMPACT SURVIVAL

If a significant asteroid fell to Earth, the first thing people would notice would be a brilliant meteor as it entered the atmosphere. The meteor would glow brighter than the Sun due to air friction, and the searing heat would scorch everything within miles around. The immense size of the asteroid would ensure that it lands virtually intact after its fiery plunge through the atmosphere.

The atmosphere would hardly slow the asteroid, which would retain 75 percent of its initial velocity. An atmospheric shock wave generated by the asteroid's blazing speed would emit a tremendous thunderclap strong enough to bowl people over within sight of the fireball. Upon impact, the explosion would produce the energy equivalent to that of a 20-megaton thermonuclear bomb, one of the most powerful nuclear devices ever built.

The impact would generate a rapidly expanding dust plume that would grow several thousand feet across at the base and rise several miles high. Most of the surrounding atmosphere would be blown away by the tremendous shock wave from the impact. A giant plume of gas and dust would punch through the atmosphere like the mushroom cloud of a nuclear bomb (Fig. 164). The rapidly rising plume would turn into an enormous dust cloud that encircles the entire Earth, turning day into dusk.

Humanity has been very fortunate not to have experienced an impact of this magnitude over the last several million years. Otherwise, evolution might have taken a decidedly different turn, and people would not be around. Impacts involving massive asteroids occur on time scales of about every 50 million years. That much time has elapsed since a large asteroid plunged into the Atlantic off Nova Scotia, producing a crater on the bottom of the ocean 30 miles wide. With this revelation in mind, Earth appears to be well overdue for another major asteroid impact.

Figure 164 *A nuclear weapon detonation at the Nevada Test Site on January 17, 1962.*

(Photo courtesy U.S. Air Force)

A collision of a large asteroid with Earth could be quite devastating. Indeed, nuclear war has been compared to the impact of a large asteroid, similar in size to the one that is supposed to have killed off the dinosaurs. The impact would send aloft huge amounts of dust into the atmosphere along with soot from global forest fires set ablaze by white-hot impact debris. The dust cloud would clog the skies and place the planet into a deep freeze for several months, a scenario also used to describe the nuclear winter effect following a full-scale nuclear war.

A large asteroid slamming into Earth would produce powerful blast waves, immense tsunamis, extremely toxic gases, and strong acid rains generated by shock heating the atmosphere. The bombardment could also inject large quantities of volcanic ash into the air from volcanoes erupting from the jolt of the impact. Likewise, earthquakes could rumble across the landscape by slipping faults set off by the impact.

If an asteroid landed in the ocean, it would form a crater on the seabed. Water rushing in to refill the crater would form a deepwater wave speeding away from the point of impact. When the wave struck the shallow waters of the continental shelf, its speed would fall rapidly and its height would rise dramatically. The wave could reach several hundred feet high and rush to shore tens of miles inland. The tsunamis created by a splashdown in the ocean would be particularly hazardous to onshore and nearshore inhabitants. If an object 1 mile wide landed in the middle of the Atlantic Ocean, for example, the resulting tsunami would wipe out coastal regions in eastern North America and western Europe.

Perhaps the worst environmental hazards would result from large volumes of suspended sediment blasted into the atmosphere. Soot from continent-sized wildfires would also clog the skies, causing darkness at noon. If the asteroid landed in the ocean, it would instantly evaporate massive quantities of seawater, saturating the atmosphere with billowing clouds of steam. This added burden would dramatically raise the density of the atmosphere and greatly increase its opacity, making penetration of sunlight nearly impossible. Sunlight unable to reach the surface would have a dramatic effect on the climate along with a significant drop in photosynthesis.

Solar radiation would also heat the darkened, sediment-laden layers, causing a thermal imbalance that could radically alter weather patterns, turning much of the land surface into a barren desert. Horrendous dust storms driven by maddening winds would rage across whole continents, further clogging the skies. Vast areas of Earth would be subjected to prolonged darkness and cold temperatures. Precipitation in the higher altitudes would cease, ensuring that the dust and smoke clouds would spread worldwide and linger for up to a year or more. Gradually, as Earth lost much of its life-sustaining properties, whole ecosystems would break down, causing mass extinctions of millions of species, thereby becoming one of the worst die-offs in Earth's history.

Agricultural crops destroyed by lack of sunlight and global cooling would lead to starvation and worldwide breakdown in the fragile human institutions of government, economics, and finally civilization itself. The environmental degradation would cause social and economic structures of all nations to collapse into total chaos. For all practical purposes, survivors would be reduced to a primitive existence. Those who survive the worst of the ordeal would be weakened by hunger, cold, and disease, resulting in high death tolls. Most of the planet would be forced to deal with a major environmental disaster reminiscent of the last days of the dinosaurs.

CONCLUSION

Although the chances of a large asteroid or comet, such as the one that killed off the dinosaurs, colliding with Earth today are highly remote, many unrecognized pieces of space debris capable of creating a crater 10 miles wide, enough to engulf a major city, are roaming freely in space. Recognition of more than 100 significant craters less than 200 million years old suggests a somewhat constant meteorite impact rate. Many craters date to within the span of time that humans have been in existence. However, people have been extremely lucky not to suffer the fates of other species that have gone extinct throughout geologic history as a result of meteorite impacts. Perhaps an unknown body somewhere out in space is taking aim at Earth, placing the planet in danger of a cosmic shooting match.

GLOSSARY

ablation (ah-BLAY-shen) the removal by vaporization of the molten surface layers of meteorites during passage through the atmosphere

abrasion erosion by friction, generally caused by rock particles carried by running water, ice, and wind

abyss (ah-BIS) the deep ocean, generally over a mile in depth

accretion (ah-KREE-shen) the accumulation of celestial dust by gravitational attraction into a planetesimal, asteroid, moon, or planet

albedo the amount of sunlight reflected from an object and dependent on color and texture

angular momentum a measure of an object or orbiting body to continue spinning

aphelion (ah-FEEL-yen) the point at which the orbit of a planet is at its farthest point from the Sun; in the case of Earth, it occurs in early July

apogee (AH-pah-gee) the point at which an object is farthest from the body it orbits

Apollo asteroids asteroids that come from the main belt between Mars and Jupiter and cross Earth's orbit

Archean (ar-KEY-an) major eon of the precambrian from 4.0 to 2.5 billion years ago

asteroid a rocky or metallic body, orbiting the Sun between Mars and Jupiter, and leftover from the formation of the solar system

asteroid belt a band of asteroids, orbiting the Sun between the orbits of Mars and Jupiter

astrobleme (AS-tra-bleem) eroded remains on the Earth's surface of an ancient impact structure produced by a large, cosmic body

axis a straight line about which a body rotates

Baltica (BAL-tik-ah) an ancient Paleozoic continent of Europe

basalt (bah-SALT) a dark, volcanic rock rich in iron and magnesium and usually quite fluid in the molten state

basement rock subterranean igneous, metamorphic, granitic, or highly deformed rock underlying younger sediments

bedrock solid layers of rock beneath younger layers

binary stars two stars closely together and orbiting each other

black hole a large, gravitationally collapsed body from which nothing, including light, can escape its powerful gravity

blue shift the shift of spectral lines toward shorter wavelengths caused by the Doppler effect for an approaching source

bolide (BULL-ide) an exploding meteor whose fireball is often accompanied by a bright light and sound when passing through Earth's atmosphere

calcareous nannoplankton planktonic plants with shells composed of calcite

calcite a mineral composed of calcium carbonate

caldera (kal-DER-eh) a large, pitlike depression at the summits of some volcanoes and formed by great explosive activity and collapse

carbonaceous (KAR-beh-NAY-shes) a substance containing carbon, namely sedimentary rocks such as limestone or certain types of meteorites

carbonaceous chondrites stony meteorites that contain abundant organic compounds

carbonate a mineral containing calcium carbonate such as limestone and dolostone

Cenozoic (sin-eh-ZOE-ik) an era of geologic time comprising the last 65 million years

Cepheid variable a star whose intensity varies periodically and used to determine distances in the universe

chondrite (KON-drite) the most common type of meteorite composed mostly of rocky material with small spherical grains called chondrules

chondrule (KON-drule) rounded granules of olivine and pyroxene found in stony meteorites called chondrites

coma (KOE-mah) the atmosphere surrounding a comet when it comes within the inner solar system; the gases and dust particles are blown outward by the solar wind to form the comet's tail.

comets   a celestial body believed to originate from a cloud of comets that surrounds the Sun; and develops a long tail of gas and dust particles when traveling near the inner solar system

continental margin   the area between the shoreline and the abyss that represents the true edge of a continent

continental shelf   the offshore area of a continent in shallow sea

continental shield   ancient crustal rocks upon which the continents grew

continental slope   the transition from the continental margin to the deep-sea basin

convection   a circular, vertical flow of a fluid medium by heating from below; as materials are heated, they become less dense and rise, cool down, and become more dense and sink

coral   a large group of shallow-water, bottom-dwelling marine invertebrates comprising reef-building colonies common in warm waters

core   the central part of a planet and that often consists of a heavy iron-nickel alloy

cosmic dust   small meteoroids existing in dust bands possibly created by the disintegration of comets

cosmic rays   high-energy charged particles that enter Earth's atmosphere from outer space

crater, meteoritic   a depression in the crust produced by the bombardment of a meteorite

crater, volcanic   the inverted conical depression found at the summit of most volcanoes, formed by the explosive emission of volcanic ejecta

craton (CRAY-ton)   the ancient, stable interior of a continent, usually composed of Precambrian rocks

Cretaceous (kri-TAE-shes)   a period of geologic time from 135 to 65 million years ago

crust   the outer layers of a planet's or moon's rocks

crustal plate   a segment of the lithosphere involved in the interaction of other plates in tectonic activity

dayglow   sunlight absorbed and reradiated by oxygen atoms in the atmosphere

diapir (DIE-ah-per)   the buoyant rise of a molten rock through heavier rock

diatom (DIE-ah-tom)   microplants whose fossil shells form siliceous sediments called diatomaceous earth

diogenite (die-OH-gin-tei)   a meteorite composed of basalt originating from the surface of a large asteroid

disk galaxy   a flat, pancake-shaped galaxy with a radius up to 25 times its thickness

Doppler effect (DAH-plir)   the effect by which the motion of the source of a wave shifts the frequency of that wave

earthquake the sudden rupture of rocks along active faults in response to geologic forces within Earth

ecliptic (ee-KLIP-tic) the plane of Earth's and other planet's orbit around the Sun

electron a negative particle of small mass orbiting the nucleus of an atom and equal in number to the protons of that atom

elliptical galaxy a galaxy whose structure is smooth and amorphous, without spiral arms, and ellipsoidal in shape

erosion the wearing away of surface materials by natural agents such as water and wind

escarpment (es-KARP-ment) a mountain wall produced by the elevation of a block of land often produced by meteorite impacts

eucrite (YUE-krite) a meteorite composed of basalt originating from the surface of a large asteroid

extraterrestrial pertaining to all phenomena outside Earth

extrusive (ik-STRU-siv) an igneous volcanic rock ejected onto the Earth's surface and other bodies in the solar system

fault a break in crustal rocks caused by earth movements

feldspar (FELL-spar) a group of rock-forming minerals comprising about 60 percent of Earth's crust and an essential component of igneous, metamorphic, and sedimentary rocks

fissure a large crack in the crust through which magma might escape from a volcano

fluvial (FLUE-vee-al) stream-deposited sediment

foraminifer(an) (FOR-eh-MI-neh-fer [un]) a calcium carbonate-secreting organism that lives in the surface waters of the oceans; after death, its shells form the primary constituent of limestone and sediments deposited onto the seafloor

formation a combination of rock units that can be traced over a distance

galaxy a large, gravitationally bound cluster of stars

gamma rays photons of very high energy and short wavelength, the most penetrating of electromagnetic radiation

geomorphology (JEE-eh-more-FAH-leh-jee) the study of surface features of Earth

geyser (GUY-sir) a spring that ejects intermittent jets of steam and hot water

glacier a thick mass of moving ice occurring where winter snowfall exceeds summer melting

glossopteris (GLOS-op-ter-is) a late Paleozoic plant that existed on the southern continents but was lacking on the northern continents, thereby confirming the existence of Gondwana

gneiss (nise) a foliated or banded metamorphic rock with similar composition as granite

Gondwana (gone-WAN-ah)   a southern supercontinent of the Paleozoic time, comprised of Africa, South America, India, Australia, Antarctica; it broke up into the present continents during the Mesozoic era

granite   a coarse-grain, silica-rich igneous rock consisting primarily of quartz and feldspars

greenstone   a green, weakly metamorphic, igneous rock

groundwater   water derived from the atmosphere that percolates and circulates below the surface

half-life   the time for one-half the atoms of a radioactive element to decay into a stable element

heliopause (HE-lee-ah-poz)   the boundary between the Sun's domain and interstellar space

helium   the second most abundant element in the universe, composed of two protons and two neutrons

hiatus (hie-AY-tes)   a break in geologic time due to a period of erosion or non-deposition of sedimentary rock, often found between geologic periods

hot spot   a volcanic center with no relation to a plate boundary; an anomalous magma generation site in the mantle

howardite (HOW-ord-ite)   a meteorite composed of basalt originating from the surface of a large asteroid

Hubble age   the approximate age of the universe obtained by extrapolating the observed expansion backward in time; the accepted value of the Hubble age is roughly 15 billion years

hydrocarbon   a molecule consisting of carbon chains with attached hydrogen atoms

hydrogen   the lightest and most abundant element in the universe, composed of one proton and one electron

hydrologic cycle   the flow of water from the ocean onto the land and back to the sea

hydrosphere   the water layer at the surface of Earth

hydrothermal   relating to the movement of hot water through the crust; it is the circulation of cold seawater downward through the oceanic crust toward the deeper depths of the oceanic crust where it becomes hot and buoyantly rises toward the surface

hypercane   a theoretical massive hurricane created by heat generated by a meteorite impact into the ocean

Iapetus Sea (eye-AP-i-tus)   a former sea that occupied a similar area as the present Atlantic Ocean prior to the assemblage of Pangaea

Icarus Project (IK-eh-rus)   a study of methods to prevent or minimize tasteroid collisions with Earth

ice age   a period of time when large areas of the Earth were covered by massive glaciers

ice cap   a polar cover of ice and snow

igneous rocks (IG-nee-es)   all rocks that have solidified from a molten state

impact   the point on the surface upon which a celestial object lands

infrared   invisible light with a wavelength between red light and radio waves

insolation   all solar radiation impinging on a planet

intrusive   a granitic body that invades Earth's crust

iridium (i-RI-dee-em)   a rare isotope of platinum, relatively abundant on meteorites and to a lesser extent on comets, but rare in Earth's crust

isostasy (eye-SOS-the-see)   a geologic principle that states that Earth's crust is buoyant and rises and sinks depending on its density

isotope (I-seh-tope)   a particular atom of an element that has the same number of electrons and protons as the other atoms of the element but a different number of neutrons; i.e., the atomic numbers are the same, but the atomic weights differ

Kirkwood gaps   bands in the asteroid belt that are mostly empty of asteroids due to Jupiter's gravitational attraction

Kuiper belt (CUE-per)   a band of comets outside the orbit of Neptune

lamellae (leh-ME-lee)   striations on the surface of crystals caused by a sudden release of high pressures such as those created by large meteorite impacts

landform   a surface feature of Earth or other planetary body

Laurasia (lure-AY-zha)   a northern supercontinent of Paleozoic time, consisting of North America, Europe, and Asia

Laurentia (lure-IN-tia)   an ancient North American continent

lava   molten magma that flows out onto the surface from a volcano

light–year   the distance that electromagnetic radiation, principally light waves travel in a vacuum in one year, or approximately 6 trillion miles

limestone   a sedimentary rock composed of calcium carbonate that is secreted from seawater by invertebrates and whose skeletons comprise the bulk of deposits

lithosphere (LI-the-sfir)   the rocky outer layer of the mantle that includes the terrestrial and oceanic crusts; the lithosphere circulates between the Earth's surface and mantle by convection currents

lithospheric plate   a segment of the lithosphere, the upper-layer plate of the mantle, involved in the interaction of other plates in tectonic activity

Magellanic Clouds   clouds of gas and dust lying outside the Milky Way galaxy

magma   a molten rock material generated within Earth and other planets and that is the constituent of igneous rocks

magnetic field reversal   a reversal of the north-south polarity of Earth's magnetic poles sometimes caused by meteorite impacts

magnetic superchron   a long duration between geomagnetic reversals, from 123 million to 83 million years ago

magnetometer   a devise that measures the intensity and direction of Earth's magnetic field

magnetosphere   the region of Earth's upper atmosphere in which the magnetic field controls the motion of ionized particles

magnitude   the relative brightness of a celestial body

mantle   the part of a planet below the crust and above the core, composed of dense rocks that might be in convective flow

maria (MAR-ee-eh)   dark plains on the lunar surface caused by massive basalt floods

mass   the measure of the amount of matter in a body

mega-tsunami   a massive wave created by a meteorite impact into the ocean

Mesozoic (MEH-zeh-ZOE-ik)   literally the period of middle life, referring to a period between 250 and 65 million years ago

metamorphism (me-the-MORE-fi-zem)   recrystallization of previous igneous, metamorphic, and sedimentary rocks under extreme temperatures and pressures without melting

meteor   a small celestial body that becomes visible as a streak of light when entering Earth's atmosphere

meteorite   a metallic or stony celestial body that enters Earth's atmosphere and impacts onto the surface

meteoritics   the science that deals with meteors and related phenomena

meteoroid   a meteor in orbit around the Sun with no relation to the phenomena it produces when entering Earth's atmosphere

meteor shower   a phenomenon observed when large numbers of meteors enter Earth's atmosphere; their luminous paths appear to diverge from a single point

microcrystals   minute mineral crystals typical of volcanic glass but absent in impact glass from meteorites

micrometeorites   small, grain-sized bodies that strike spacecraft in orbit around Earth or in outer space

microtektites   small, spherical grains created by the melting of surface rocks during a large meteorite impact

near-Earth asteroid   an asteroid outside the asteroid belt in an Earth-crossing orbit

nebula (NEH-by-lah)   an extended astronomical object with a cloudlike appearance; some nebulae are galaxies, others are clouds of dust and gas within the Milky Way galaxy

Nemesis (NEE-mah-sez)   the hypothetical sister star of the Sun, responsible for disturbing comets in the Oort cloud and causing them to fall into the inner solar system

neutron a particle with no electrical charge and that has roughly the same weight as the positively charged proton, both of which are found in the nucleus of an atom

nova (NOE-vah)   a star that suddenly brightens during its final stages

olivine a greenish, rock-forming mineral in Earth's interior and on lunar rocks

Oort cloud (ort)   the collection of comets that surround the Sun about a light-year away

ophiolite (oh-FI-ah-lite)   oceanic crust thrust upon continents by plate tectonics

orbit the circular or elliptical path of one body around another

outgassing the loss of gas within a planet as opposed to degassing, or loss of gas from meteorites

ozone a molecule consisting of three atoms of oxygen in the upper atmosphere and that filters out ultraviolet radiation from the Sun

paleomagnetism the study of Earth's magnetic field, including the position and polarity of the poles in the past to determine the position of the continents

Paleozoic (PAY-lee-eh-ZOE-ik)   the period of ancient life, between 540 and 250 million years ago

Pangaea (pan-GEE-a)   a Paleozoic supercontinent that included all the lands of the Earth

Panthalassa (pan-THE-lass-ah)   the global ocean that surrounded Pangaea during the Mesozic

peridotite (pah-RI-deh-tite)   the most common rock type in the mantle; also found in lunar rocks

perigee (PER-eh-gee)   the point at which an object is closest to the body it orbits

perihelion (per-eh-HEEL-yen)   the point at which the orbit of a planet is at its nearest to the Sun; in the case of Earth, it occurs in early January

planetesimals (pla-neh-THE-seh-mals)   small celestial bodies that might have existed in the early stages of the solar system and coalesced to form the planets

planetoid (PLA-neh-toid)   a small body generally no larger than the Moon in orbit around the Sun; a disintegration of several such bodies might have been responsible for the asteroid belt between Mars and Jupiter.

plate tectonics   the theory that accounts for the major features of Earth's surface in terms of the interaction of lithospheric plates; some other planets also might exhibit this feature

precession   the slow change in direction of Earth's axis of rotation due to gravitational action of the Moon on Earth

Proterozoic   a period of geologic history between 2.5 and 0.6 billion years ago

proton   a large particle with a positive charge in the nucleus of an atom

pulsar   a highly energetic, starlike object that radiates intense radio signals

pyroxene (pie-ROCK-seen)   a black, rock-forming mineral in Earth's interior and in lunar rocks

quasar (KWAY-zar)   a quasi-stellar object that is considered the brightest and most distant of objects in the universe used to determine its size

radiant   a point in space from which the luminous tails of meteors appear to diverge during a meteor shower; they appear to come from a central source

radiolarian (RAY-dee-oh-LAR-eh-en)   a microorganism with shells made of silica comprising a large component of siliceous sediments

radiometric dating   determining the age of an object by radiometrically and chemically analyzing its stable and unstable radioactive elements. By this method, meteorites established the age of Earth

radionuclide (RAY-dee-oh-NEW-klide)   a radioactive element that is responsible for generating Earth's internal heat as well as other bodies in the solar system

red shift   the shift of light toward the lower end of the spectrum indicating that distant galaxies are receding, proving the age of the universe

regolith (REE-geh-lith)   loose rock material on the Moon's surface

retrograde (REH-treh-grade)   a rotation or revolution in the opposite direction of the other bodies in the solar system. Many moons in the outer solar system exhibit this phenomenon

revolution   the motion of a celestial body in its orbit as with Earth around the Sun or with asteroids and comets circling the solar system

rille (ril)   a trench formed by a collapsed lava tunnel on Earth, the Moon, and other planets

riverine (RI-vah-rene)   relating to a river

rotation   the turning of a body about its axis; as with Earth the rotation rate is 24 hours

sandstone   a sedimentary rock consisting of sand grains cemented together

schist (shist)   a finely layered metamorphic, crystalline rock easily split along parallel bands

seismic (SIZE-mik)   pertaining to earthquake energy or other violent ground vibrations sometimes caused by meteorite impacts

shield   areas of exposed Precambrian nucleus of a continent where many meteorite craters have been found

siderite (SI-dir-ite)   a nickel-iron meteorite

siderophiles (si-DER-eh-phile)   literally meaning "iron lovers," elements that combine with iron and are carried down to Earth's core during meteorite bombardment

sinkhole a large pit formed by the collapse of surface materials undercut by the dissolution of subterranean limestone. Such sinkholes outlined the Chicxulub crater, proving its existence

solar wind an outflow of particles from the Sun, which represents the expansion of the corona; the reaction of the solar wind on cometary nuclei along with solar heating produces the tails of comets

spherules (SFIR-ule) small, spherical, glassy grains found on certain types of meteorites, on lunar soils, and at large meteorite impact sites

spiral galaxy a galaxy, like the Milky Way, with a prominent central bulge embedded in a flat disk of gas, dust, and young stars that wind out in spiral arms from the nucleus

stishovite (STIS-hoe-vite) a quartz mineral produced by extremely high pressures such as those generated by a large meteorite impact

stratigraphy the study of rock strata; used to determine the age of meteorite impacts

strewn field an usually large area where tektites are found arising from a large meteorite impact

striae (STRY-aye) scratches on bedrock made by rocks embedded in a moving glacier

subduction zone a region where an oceanic plate dives below a continental plate into Earth's mantle; ocean trenches are the surface expression of a subduction zone. No other planet exhibits this phenomenon except possibly the moons of the outer planets.

supercluster a large association of galaxies that exist far outside this galaxy

supernova an enormous stellar explosion in which all but the inner core of a star is blown off into interstellar space, producing as much energy in a few days as the Sun does in a billion years

tectonics (tek-TAH-nik) the history of Earth's larger features (rock formations and plates) and the forces and movements that produce them; this feature might also exist on other planets and moons

tektites (TEK-tites) small, glassy minerals created from the melting of surface rocks by an impact of a large meteorite

terrestrial all phenomena pertaining to Earth, including meteorite impacts

Tethys Sea (THE-this) the hypothetical, midlatitude region of the oceans separating the northern and southern continents of Laurasia and Gondwana several hundred million years ago

tsunami (sue-NAH-me) a sea wave generated by an undersea earthquake or marine volcanic eruption; a large meteorite landing into the ocean would produce a tremendous tsunami

T-Tauri wind a strong particle radiation from newly formed stars

tundra permanently frozen ground at high latitudes; most craters are well preserved in these regions

ultraviolet the invisible light with a wavelength shorter than visible light and longer than X rays

volcanism any type of volcanic activity, including that on other planets and moons

volcano a fissure or vent in the crust through which molten rock rises to the surface to form a mountain

X rays electromagnetic radiation of high-energy wavelengths above the ultraviolet and below the gamma rays

BIBLIOGRAPHY

ORIGIN OF THE SOLAR SYSTEM

Bothun, Gregory D. "The Ghostliest Galaxies." *Scientific American* 276 (February 1997): 56–61.

Cowen, Ron. "Revved-Up Universe." *Science News* 150 (February 12, 2000): 286–287.

Freedman, David H. "The Mediocre Universe." *Discover* 17 (February 1996): 65–75.

Glanz, James. "CO in the Early Universe Clouds Cosmologists' Views." *Science* 273 (August 2, 1996): 581.

Gorman, Jessica. "Cosmic Chemistry Gets Creative." *Science News* 159 (May 19, 2001): 317–319.

Hogan, Craig J. "Primordial Deuterium and the Big Bang." *Scientific American* 275 (December 1996): 68–73.

Horgan, John. "Pinning Down Inflation." *Scientific American* 276 (June 1997): 17–18.

Macchetto, F. Duccio, and Mark Dickinson. "Galaxies in the Young Universe." *Scientific American* 276 (May 1997): 93–99.

Sutton, Christine. "How To Stop a Supernova Stalling." *New Scientist* 144 (November 19, 1994): 20–21.

Veilleux, Sylvain, Gerald Cecil, and Jonathan Bland-Hawthorn. "Colossal Galactic Explosions." *Scientific American* 274 (February 1996): 98–103.

Write, Ian. "Ingrained Evidence of Origin." *Nature* 365 (October 28, 1993): 786–787.

THE FORMATION OF EARTH

Allegre, Claude J., and Stephen H. Schnider. "The Evolution of the Earth." *Scientific American* 271 (October 1994): 66–75.

Cowen, Ron. "Water Flowed Early in the Solar System." *Science News* 149 (February 24, 1996): 117.

Halliday, Alex N., and Michael J. Drake. "Colliding Theories." *Science* 283 (March 19, 1999): 1861–1864.

Holland, Heinrich D. "Evidence for Life on Earth More Than 3850 Million Years Ago." *Science* 275 (January 3, 1997): 38–39.

Horgan, John. "In the Beginning." *Scientific American* 264 (February 1991): 117–125.

Kasting, James F. "New Spin on Ancient Climate." *Nature* 364 (August 26, 1993): 759–761.

Marcus, Joseph. "Did a Rain of Comets Nurture Life?" *Science* 254 (November 22, 1991): 1110–1111.

Newsom, Horton E., and Kenneth W. Sims. "Core Formation During Early Accretion of the Earth." *Science* 252 (May 17, 1991): 926–933.

Orgel, Leslie E. "The Origin of Life on the Earth." *Scientific American* 271 (October 1994): 77–83.

Powell, Corey S. "Peering Inward." *Scientific American* 264 (June 1991): 101–111.

Taylor, S. Ross, and Scott M. McLennan. "The Evolution of Continental Crust." *Scientific American* 274 (January 1996): 76–81.

Vogel, Shawna. "Living Planet." *Earth* 5 (April 1996): 27–35.

CRATERING EVENTS

Dalziel, Ian W. D. "Earth Before Pangaea." *Scientific American* 272 (January 1995): 58–63.

Hildebrans, Alan R., and William V. Boynton. "Cretaceous Ground Zero." *Natural History* 104 (June 1991): 47–52.

Hoffman, Paul F., and David P. Schrag. "Snowball Earth." *Scientific American* 282 (January 2000): 68–75.

Kerr, Richard A. "Geologists Pursue Solar System's Oldest Relics." *Science* 290 (December 22, 2000): 2239–2242.

———. "Making an Impact Under the Chesapeake." *Science* 265 (August 19, 1994): 1036.

Knoll, Andrew H. "End of the Proterozoic Eon." *Scientific American* 265 (October 1991): 64–73.

Lowe, Donald R., et al. "Geological and Geochemical Record of 3400-Million-Year-Old Terrestrial Meteorite Impacts." *Science* 245 (September 1989): 959–962.

Moores, Eldridge. "The Story of Earth." *Earth* 5 (December 1996): 30–33.

Naeye, Robert. "The Hole in Nebraska." *Discover* 14 (April 1993): 18.

Schueller, Gretel. "The Splash Felt 'Round the World." *Earth* 7 (April 1998): 12–13.

York, Derek. "The Earliest History of the Earth." *Scientific American* 268 (January 1993): 90–96.

PLANETARY IMPACTS

Bullock, Mark A., and David H. Grinspoon. "Global Climate Change on Venus." *Scientific American* 280, (March 1999): 50–57.

Carroll, Michael. "Assault on the Red Planet." *Popular Science* 250 (January 1997): 44–49.

Cowen, Ron. "Martian Highlands: Clues to a Watery Past." *Science News* 144 (June 5, 1993): 357.

Johnson, Torrence V. "The *Galileo* Mission to Jupiter and Its Moons." *Scientific American* 282 (February 2000): 40–54.

Johnson, Torrence V., Robert Hamilton Brown, and Laurence A. Soderblom. "The Moons of Uranus." *Scientific American* 256 (April 1987): 48–60.

Kaula, William M. "Venus: A Contrast in Evolution to Earth." *Science* 247 (March 9, 1990): 1191–1196.

Kerr, Richard A. "*Galileo* Turns Geology Upside Down on Jupiter's Icy Moons." *Science* 274 (October 18, 1996): 341.

———. "The Solar System's New Diversity." *Science* 265 (September 2, 1994): 1360–1362.

Nelson, Robert M. "Mercury: The Forgotten Planet." *Scientific American* 277 (November 1997): 56–67.

Pappalardo, Robert T., James W. Head, and Ronald Greeley. "The Hidden Ocean of Europa." *Scientific American* 281 (October 1999): 54–63.

Taylor, G. Jeffrey. "The Scientific Legacy of Apollo." *Scientific American* 271 (July 1994): 40–47.

ASTEROIDS

Binzel, Richard P., M. Antonietta, and Marcello Fulchignoni. "The Origins of the Asteroids." *Scientific American* 265 (October 1991): 88–94.

Cowen, Ron. "Asteroids Get Solar Push Toward Earth." *Science News* 155 (March 6, 1999): 151.

———. "After the Fall." *Science News* 148 (October 14, 1995): 248–249.

Gehrels, Tom. "Asteroids and Comets." *Physics Today* 38 (February 1985): 33–41.

Gibbs, W. Wayt. "The Search for Greenland's Mysterious Meteor." *Scientific American* 279 (November 1998): 72–79.

Kerr, Richard A. "*NEAR* Finds a Battered but Unbroken Eros." *Science* 287 (February 25, 2000): 1378–1379.

———. "Small Asteroids Point to Source for Meteorites." *Science* 285 (August 13, 1999): 1002–1003.

Naeye, Robert. "A Ring Around the Sun." *Discover* 15 (November 1994): 31.

Perth, Nigel Henbest. "Meteorite Bonanza in Australian Desert." *New Scientist* 129 (April 20, 1991): 20.

Roach, Mary. "Meteorite Hunters. *Discover* 18 (May 1997): 71–75.

Rubin, Alan E. "Fragments of History Preserved." *Nature* 368 (April 21 1994): 691

COMETS

Cowen, Ron. "A Comet's Odd Orbit Hints at Hidden Planet." *Science News* 159 (April 7, 2001): 313.

———. "Comets: Mudballs of the Solar System?" *Science News* 141 (March 14, 1992): 170–171.

DiChristina, Mariette. "Lessons from Hale-Bopp." *Popular Science* 238 (August 1997): 65–66.

Hecht, Jeff. "All Eyes on Jupiter's Collision." *New Scientist* 134 (November 6, 1993): 18–19.

Horgan, John. "Beyond Neptune." *Scientific American* 273 (October 1995): 24–25.

Kerr, Richard A. "Shoemaker-Levy Dazzles, Bewilders." *Science* 265 (July 29, 1994): 601–602.

Luu, Jane X., and David C. Jewitt. "The Kuiper Belt." *Scientific American* 274 (May 1996): 46–52.

Monastersky, Richard. "Is the Earth Pelted by Space Snowballs?" *Science News* 151 (May 31, 1997): 332.

Seofe, Charles. "Do Comets Get a Nudge from the Galaxy?" *Science* 274 (November 8, 1996): 920.

Weissman, Paul R. "The Oort Cloud." *Scientific American* 279 (September 1998): 84–89.
———. "Bodies on the Brink." *Nature* 374 (April 27, 1995): 762.
Yamamoto, Tetsuo. "Are Edgeworth-Kuiper Belt Objects Pristine?" *Science* 273 (August 16, 1996): 921.

METEORITE CRATERS

Grieve, Richard A. F. "Impact Cratering on the Earth." *Scientific American* 262 (April 1990): 66–73.
Kerr, Richard A. "Testing an Ancient Impact's Punch." *Science* 263 (March 11, 1994): 1371–1372.
———. "Huge Impact Tied to Mass Extinction." *Science* 257 (August 14, 1992): 878–880.
Long, Michael E. "Mars on Earth." *National Geographic* 196 (July 1999): 34–51.
Melosh, H. J. "Around and Around We Go. *Nature* 376 (August 3, 1995): 386–387.
———. "Under the Ringed Basins." *Nature* 373 (January 12, 1995): 104–105.
Monastersky, Richard. "Target Earth." *Science News* 153 (May 16, 1998): 312–314.
———. "Shots from Outer Space." *Science News* 147 (January 28, 1995): 58–59.
O'Keffe, John A. "The Tektite Problem." *Scientific American* 239 (August 1978): 116–125.
Sharpton, Virgil L. "Glasses Sharpen Impact Views." *Geotimes* 33 (June 1988): 10–11.
Smit, Jan. "Where Did It Happen?" *Nature* 349 (February 7, 1991): 641–642.

IMPACT EFFECTS

Alvarez, Luis W. "Mass Extinctions Caused by Large Bolide Impacts." *Physics Today* 40 (July 1987): 24–33.
Erwin, Douglas H. "The Mother of Mass Extinctions." *Scientific American* 275 (July 1996): 72–78.
Fields, Scott. "Dead Again." *Earth* 4 (April 1995): 16.
Fischman, Joshua. "Flipping the Field." *Discover* 11 (May 1990): 28–29.
Kerr, Richard A. "Earth's Core Spins at Its Own Rate." *Science* 273 (July 26, 1996): 428–429.

———. "A Bigger Death Knell for the Dinosaurs?" *Science* 261 (September 17, 1993): 1518–1519.

Monastersky, Richard. "Cretaceous Extinctions: The Strikes Add Up." *Science News* 144 (June 19, 1993): 391.

Morell, Virginia. "How Lethal Was the K–T Impact?" *Science* 261 (September 17, 1993): 1518–1519.

Pendick, Daniel. "The Dust Ages." *Earth* 5 (June 1996): 22–23, 66.

Schwarzchild, Bertram. "Do Asteroid Impacts Trigger Geomagnetic Reversals?" *Physics Today* 40 (February 1987): 17–20.

Svitil, Kathy A. "Hurricane from Hell." *Discover* 17 (April 1995): 26.

Zimmer, Carl. "Inconstant Field." *Discover* 15 (February 1994): 26–27.

DEATH STAR

Beardsley, Tim. "Star-Struck?" *Scientific American* 258 (April 1988): 37–40.

Benton, Michael J. "Late Triassic Extinctions and the Origin of the Dinosaurs." *Science* 260 (May 7, 1993): 769–770.

Cowen, Ron. "Spotlight on Betelgeuse." *Science News* 149 (January 27, 1996): 63.

Goldsmith, Donald. "Supernova Offers a First Glimpse of Universe's Fate." *Science* 276 (April 4, 1997): 37–38.

Hildebrand, Alan R., and William V. Boynton. "Cretaceous Ground Zero." *Natural History* (June 1991): 47–52.

Kerr, Richard A. "Extinction by a One-Two Comet Punch." *Science* 255 (January 10, 1992): 160–161.

Monastersky, Richard. "Cretaceous Die-Offs: A Tale of Two Comets?" *Science News* 143 (April 3, 1993): 212–213.

———. "Closing in on the Killer." *Science News* 141 (January 25, 1992): 56–58.

Waters, Tom. "The Dinosaur Acid Test." *Discover* 11 (February 1990): 28.

Weissman, Paul R. "Are Periodic Bombardments Real?" *Sky & Telescope* 79 (March 1990): 266–270.

COSMIC COLLISIONS

Alvarez, Walter, and Frank Asaro. "An Extraterrestrial Impact." *Scientific American* 263 (October 1990): 78–84.

Cowen, Ron. "Rocky Relics." *Science News* 145 (February 5, 1994): 88–90.

Desonie, Dana. "The Threat from Space." *Earth* 5 (August 1996): 25–31.

Gehrels, Tom. "Collisions with Comets and Asteroids." *Scientific American* 274 (March 1996): 54–59.

Irion, Robert. "Asteroid Searchers Streak Ahead." *Science* 281 (August 14, 1998): 894–895.

Mathews, Robert. "A Rocky Watch for Earthbound Asteroids." *Science* 255 (March 6, 1992): 1204–1205.

Morrison, David. "Target Earth: It Will Happen." *Sky & Telescope* 79 (March 1990): 261–265.

Peterson, Ivars. "Tunguska: The Explosion of a Stony Asteroid." *Science News* 143 (January 9, 1993): 23.

Powell, Corey S. "Asteroid Hunters." *Scientific American* 268 (April 1993): 34–40.

Sinnott, Roger W. "An Asteroid Whizzes Past Earth." *Sky & Telescope* 78 (July 1989): 30.

Stone, Richard. "The Last Great Impact on Earth." *Discover* 17 (September 1996): 60–71.

Yam Philip. "Making a Deep Impact." *Scientific American* 278 (May 1998): 22–24.

INDEX

Boldface page numbers indicate extensive treatment of a topic. *Italic* page numbers indicate illustrations or captions. Page numbers followed by *m* indicate maps; *t* indicate tables; *g* indicate glossary.